AF360821

Advances in Multiscale and Multifield Solid Material Interfaces

Editors

Raffaella Rizzoni
Frédéric Lebon
Serge Dumont
Michele Serpilli

MDPI • Basel • Beijing • Wuhan • Barcelona • Belgrade • Manchester • Tokyo • Cluj • Tianjin

Editors
Raffaella Rizzoni
Department of Engineering
University of Ferrara
Ferrara
Italy

Frédéric Lebon
CNRS, Laboratoire de
Mécanique et d'Acoustique
Université Aix-Marseille
Marseille
France

Serge Dumont
Laboratoire IMAG
Montpellier
Université de Nîmes
Nîmes
France

Michele Serpilli
Department of Civil and
Building Engineering, and
Architecture
Università Politecnica delle
Marche
Ancona AN
Italy

Editorial Office
MDPI
St. Alban-Anlage 66
4052 Basel, Switzerland

This is a reprint of articles from the Special Issue published online in the open access journal *Technologies* (ISSN 2227-7080) (available at: www.mdpi.com/journal/technologies/special_issues/ Solid_Material_Interfaces).

For citation purposes, cite each article independently as indicated on the article page online and as indicated below:

LastName, A.A.; LastName, B.B.; LastName, C.C. Article Title. *Journal Name* **Year**, *Volume Number*, Page Range.

ISBN 978-3-0365-3414-5 (Hbk)
ISBN 978-3-0365-3413-8 (PDF)

Contents

 technologies

Editorial

Editorial for the Special Issue "Advances in Multiscale and Multifield Solid Material Interfaces"

Raffaella Rizzoni [1,*,†], **Frédéric Lebon** [2,†], **Serge Dumont** [3,†] and **Michele Serpilli** [4,†]

1   Department of Engineering, University of Ferrara, 44122 Ferrara, Italy
2   CNRS, Centrale Marseille, Laboratoire de Mécanique et d'Acoustique, Aix-Marseille University, 13453 Marseille, France; lebon@lma.cnrs-mrs.fr
3   IMAG CNRS UMR 5149, University of Nîmes, 30000 Nîmes, France; serge.dumont@unimes.fr
4   Department of Civil and Building Engineering, and Architecture, Università Politecnica delle Marche, 60121 Ancona, Italy; m.serpilli@univpm.it
*   Correspondence: raffaella.rizzoni@unife.it; Tel.: +39-0532-974959
†   These authors contributed equally to this work.

Citation: Rizzoni, R.; Lebon F.; Dumont, S.; Serpilli, M. Editorial for the Special Issue "Advances in Multiscale and Multifield Solid Material Interfaces". *Technologies* **2022**, *10*, 34. https://doi.org/10.3390/technologies10010034

Received: 9 February 2022
Accepted: 11 February 2022
Published: 18 February 2022

Publisher's Note: MDPI stays neutral with regard to jurisdictional claims in published maps and institutional affiliations.

Interfaces play an essential role in determining the mechanical properties and the structural integrity of a wide variety of technological materials. As new manufacturing methods become available, interface engineering and architecture at multiscale length levels in multi-physics materials open up to applications with high innovation potential. This Special Issue is dedicated to recent advances in fundamental and applications of solid material interfaces. It contains six high-quality articles that were accepted after a careful reviewing process.

Micromechanical models for multiphase composites are fundamental for an accurate design and optimization of engineering materials, in order to predict their effective material properties and give insight into the relation between microstructure and macroscopic mechanical behavior. Three papers of the Special Issue push the boundary to reveal new insights on micromechanical models for composites incorporating interface effects [1–3].

In [1], Nazarenko et al. propose a mathematical model employing the concept of energy-equivalent inhomogeneity to analyze short cylindrical fiber composites with interfaces described by the Steigmann–Ogden material surface model. Closed-form expressions for components of the stiffness tensor of equivalent fiber have been developed and, in the limit, they have shown to compare well with the results available in the literature for infinite fibers with the Steigmann–Ogden interface model.

In [2], Rudoy considers an equilibrium problem of the Kirchhoff–Love plate containing a nonhomogeneous inclusion. The elastic properties of the inclusion rescale as ε^N, with $N < 1$ and ε a small parameter characterizing the width of the inclusion. The passage to the limit as the parameter ε tends to zero is justified, and an asymptotic model of a plate containing a thin inhomogeneous hard inclusion is constructed. It is shown that there exists two types of thin inclusions: rigid inclusion ($N < -1$) and elastic inclusion ($N = -1$). The inhomogeneity is shown to disappear in the case of $N \in (-1, 1)$.

The paper by Sabina et al. [3] implements a two-scale asymptotic homogenization method to calculate the out-of-plane effective complex-value properties of periodic three-phase elastic fiber-reinforced composites with parallelogram unit cells. Matrix and inclusions materials have complex-valued properties. Closed analytical expressions for the local problems and the out-of-plane shear effective coefficients are given. The solution of the homogenized local problems is found using potential theory. Numerical results are reported and comparisons with data in the literature show good agreement.

The mathematical treatment of contact problems for engineering applications is typically very challenging. In [4], Sofonea and Shillor propose the application of the Tykhonov well-posedness concept, which allows a unified and elegant framework for a class of static contact problems. In particular, they present an original unified approach to the analysis of contact problems with various interface laws modeling the contact between a deformable body and a rigid or reactive foundation. A weak formulation of the equilibrium problem is derived, which is in the form of an elliptic variational inequality, and the Tykhonov well-posedness of the problem is established, under appropriate assumptions on the data and parameters, with respect to a special Tykhonov triple. This abstract result leads to different convergence results, which establish the continuous dependence of the weak solution on the data and the parameters. The work enables to elucidate the links among the weak solutions of the different models and their corresponding mechanical interpretations.

In recent decades, adhesive bonded technology has been increasingly used for reduction of structural weight, time, and manufacturing costs, also due to improved mechanical performance and better understanding of failure mechanics. To assess the structural integrity of the joint, an estimate of stress distribution and a suitable failure criterion are necessary. Although for complex geometries and elaborate material models finite element analysis are available, analytical models giving closed-form results are more appropriate. In [5], Raffa et al. present a new analytical model for thin structural adhesives in glued tube-to-tube butt joints. A nonlinear and rate-dependent imperfect interface law is proposed, able to accurately describe brittle and ductile stress-strain behaviors of adhesive layers under combined tensile-torsion loads and explicitly accounting for material and damage properties of the adhesive layer. A first comparison with experimental data available in the literature provides promising results in terms of the reproducibility of the stress–strain behavior for pure tensile and torsional loads.

Structural composite materials are nowadays being engineered employing multiphysics materials to achieve superior functionalized properties. The work by Serpilli et al. [6] proposes new interface conditions between the layers of a three-dimensional composite structure in the framework of coupled thermoelasticity. More precisely, the mechanical behavior of two linear isotropic thermoelastic solids, bonded together by a thin layer, constituted of a linear isotropic thermoelastic material, is studied by means of an asymptotic analysis. After defining a small vanishing parameter ε associated with the thickness and the constitutive coefficients of the intermediate layer, two different limit models and their associated limit problems are characterized, the so-called soft and hard thermoelastic interface models. A numerical example is presented to show the efficiency of the proposed methodology, based on a finite element approach developed previously.

The papers published in this Special Issue constitute only a further step to advance in the field of multiscale and multifield solid material interfaces. However, they extend the frontiers of what researchers are already working on and will continue to investigate.

Acknowledgments: The guest editors thank the authors for their invaluable scientific contributions to the Special Issue and the anonymous reviewers for their effort and expertise that they contribute to reviewing, which were fundamental for maintaining a high scientific standard. The Special Issue is published under the auspices of the Italian National Group of Mathematical Physics (GNFM-INdAM).

Conflicts of Interest: The authors declare no conflict of interest.

References

1. Nazarenko, L.; Stolarski, H.; Altenbach, A. Modeling Cylindrical Inhomogeneity of Finite Length with Steigmann-Ogden Interface. *Technologies* **2020**, *8*, 78. [CrossRef]
2. Rudoy, E. Asymptotic Justification of Models of Plates Containing Inside Hard Thin Inclusions. *Technologies* **2020**, *8*, 59. [CrossRef]
3. Sabina, F.J.; Espinosa-Almeyda, Y.; Guinovart-Díaz, R.; Rodríguez-Ramos, R.; Camacho-Montes, H. Effective Complex Properties for Three-Phase Elastic Fiber-Reinforced Composites with Different Unit Cells. *Technologies* **2021**, *9*, 12. [CrossRef]
4. Sofonea, M.; Shillor, M. Tykhonov Well-Posedness and Convergence Results for Contact Problems with Unilateral Constraints. *Technologies* **2021**, *9*, 1. [CrossRef]
5. Raffa, M.L.; Rizzoni, R.; Lebon, F. A Model of Damage for Brittle and Ductile Adhesives in Glued Butt Joints. *Technologies* **2021**, *9*, 19. [CrossRef]
6. Serpilli, M.; Dumont, S.; Rizzoni, R.; Lebon, F. Interface Models in Coupled Thermoelasticity. *Technologies* **2021**, *9*, 17. [CrossRef]

 *technologies*

Article

Modeling Cylindrical Inhomogeneity of Finite Length with Steigmann–Ogden Interface

Lidiia Nazarenko [1,*], Henryk Stolarski [2] and Holm Altebach [1]

[1] Institute of Mechanics, Otto von Guericke University Magdeburg, Universitätsplatz 2, 39016 Magdeburg, Germany; holm.altenbach@ovgu.de

[2] Department of Civil, Environmental and Geo-Engineering, University of Minnesota, 500 Pillsbury Drive S.E., Minneapolis, MN 55455, USA; stola001@umn.edu

* Correspondence: lidiia.nazarenko@ovgu.de

Received: 30 November 2020; Accepted: 15 December 2020; Published: 18 December 2020

Abstract: A mathematical model employing the concept of energy-equivalent inhomogeneity is applied to analyze short cylindrical fiber composites with interfaces described by the Steigmann–Ogden material surface model. Real inhomogeneity consists of a cylindrical fiber of finite length, and its surface possessing different properties is replaced by a homogeneous, energy-equivalent cylinder. The properties of the energy-equivalent fiber, incorporating properties of the original fiber and its interface, are determined on the basis of Hill's energy equivalence principle. Closed-form expressions for components of the stiffness tensor of equivalent fiber have been developed and, in the limit, shown to compare well with the results available in the literature for infinite fibers with the Steigmann–Ogden interface model. Dependence of those components on the radius, length of the cylindrical fiber, and surface parameters is included in these expressions. The effective stiffness tensor of the short-fiber composites with so-defined equivalent cylindrical fibers can be determined by any homogenization method developed without accounting for interface.

Keywords: equivalent cylinder of finite length; Steigmann–Ogden surface model; anisotropic properties

---

1. Introduction

Interphases between inhomogeneities and the matrix may have a very pronounced influence on the effective behavior of entire composites. The interphases are typically three-dimensional continua, but treating them as such is feasible only for simple geometry of the inhomogeneities and for simple loading conditions.

To cover more complex situations, some effort has been invested into developing various simplified models of interphases [1–11], among others. The most practical and popular of them are the Gurtin–Murdoch [12,13] model and the spring layer model [1,4,5,14–19]. The former is a membrane-type model in which the bending stiffness of the interphase is assumed to be negligible and which preserves kinematic continuity. The latter allows for displacement discontinuity and relates forces transmitted across the interphase to the tangential and normal components of that discontinuity. The Gurtin–Murdoch and related models of surface elasticity have been used to study beams, plates, and shells [20,21].

Generalization of the Gurtin–Murdoch model was proposed by Steigmann and Ogden [22,23], who introduced the resistance of the surface to both stretching and bending. Their development is based on the Kirchhoff–Love shell kinematics and, as such, implies that the surface energy in the Steigmann–Ogden model includes both the surface membrane strain tensor and the surface curvature tensor. The Steigmann–Ogden model was used in [24,25] to study bending of nano-sized cantilever beams. In these investigations, the Steigmann–Ogden constants were determined by using a combination of atomistic simulations and a simple continuum model. A similar analysis within the Steigmann–Ogden model but for laminates was provided in [26], where the formula for the effective bending stiffness and its dependence on the surface elastic moduli were derived.

In [27–30], it was demonstrated that higher-gradient theories could entail surface tensors of stresses and couple stresses as well as other stress resultants.

Within Toupin–Mindlin formulation [31–33] of the strain gradient elasticity, the mathematical study of static and dynamic boundary value problems with surface stresses described by Steigmann-Ogden model was presented in [34,35].

The boundary conditions for the Steigmann–Ogden [22,23] model for a two-dimensional surface using general expression for surface energy were re-derived in [36]. The effect of curvature-dependent interfacial energy was also studied in [37] for finite deformation. The effective moduli of nanocomposites were analyzed in [38]. The effective properties of the isotropic particulate composites with Steigmann–Ogden interface were derived in [39,40].

In this work, the energy-equivalent inhomogeneity (EEI) approach, recently presented in [19,41–43], is applied to short fibers modeled as cylindrical inhomogeneity of finite length with a Steigmann–Ogden model of interface. The presented approach can be used for determination of the properties of equivalent homogeneous cylindrical fiber for which the properties incorporate properties of the interface and then, in combination with any homogenization method developed for composites without interfaces, for determination of effective properties of short-fiber composites with an interface.

This paper is organized as follows. The next section briefly introduces the notion of energy equivalence and its subsequent specification for cylinders of finite length and the Steigmann–Ogden model of interface; it also defines the properties of the energy-equivalent cylinder. In Section 3, this is followed by a comparison with the results available in the literature for infinite cylindrical fibers with Gurtin–Murdoch and Steigmann–Ogden interface models. The paper final section contains some overall comments about the approach pursued herein and the results obtained. Several technical details are presented in Appendix A.

2. Energy-Equivalent Short Cylindrical Fiber with Steigmann–Ogden Surface Model

2.1. General Considerations

To find properties of the equivalent inhomogeneity of any shape meant to incorporate properties of the original inhomogeneity and those of its interphase, the system is subjected to boundary displacements consistent with constant straining, represented by an arbitrary constant tensor ε_{eq}. The elastic energy of this system is

$$E = \frac{1}{2} \int_{V_1} \varepsilon_1 : \mathbf{C}_1 : \varepsilon_1 dV_1 + E_{int},\tag{1}$$

where E_{int} is the strain energy of the interphase appropriate for the Steigmann–Ogden model, ε_1 is the strain within the original inhomogeneity caused by ε_{eq}, and $\mathbf{C}_1$ is the rank four tensor of the elastic moduli of the original cylindrical inhomogeneity (Figure 1).

The mathematical description of energy equivalence is expressed by the following equation:

$$E = \frac{1}{2} \int_{V_{eq}} \varepsilon_{eq} : \mathbf{C}_{eq} : \varepsilon_{eq} dV_{eq} = \frac{1}{2} V_{eq} \varepsilon_{eq} : \mathbf{C}_{eq} : \varepsilon_{eq} = E_{int} + \frac{1}{2} \int_{V_1} \varepsilon_1 : \mathbf{C}_1 : \varepsilon_1 dV_1,\tag{2}$$

where $\mathbf{C}_{eq}$ is the unknown constitutive tensors of the equivalent inhomogeneity and E_{int} depends on the specific model of the interphase employed and on the data characterizing the system. Under the assumption of linearly elastic interphase, at equilibrium, both terms on the far right-hand side are quadratic functions of ε_{eq} and Equation (2) can be used to determine $\mathbf{C}_{eq}$. As shown in [19,41], that simple idea may be technically quite demanding, particularly for complex shapes of inhomogeneity, but it is executable and, in the cases considered so far, leads to remarkably accurate, closed-form results.

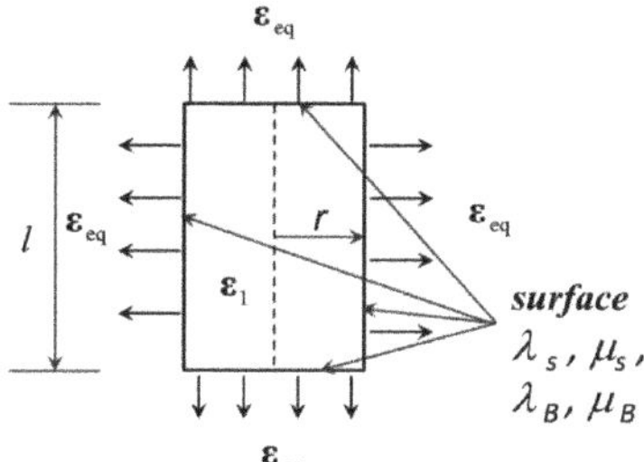

Figure 1. Schematic illustration of cylindrical inhomogeneity.

2.2. Steigmann–Ogden Surface Model and Associated Elastic Energy

The original development of the Steigmann–Ogden model, including equilibrium equations and related boundary conditions, are presented in [22,23]. These relations are derived within the Toupin–Mindlin formulation of the strain gradient elasticity in [35] and have following forms:

Displacements continuity on S_I

$$[\mathbf{u}(\mathbf{x})]_{S_I} = 0, \tag{3}$$

Stress discontinuity on S_I

$$[\boldsymbol{\sigma}(\mathbf{x})]_{S_I} \cdot \mathbf{n}(\mathbf{x}) = \nabla_{S_I} \cdot \left[\boldsymbol{\sigma}_S(\mathbf{x}) - (\nabla_{S_I} \cdot \mathbf{M}_S(\mathbf{x}))\mathbf{n}(\mathbf{x}) \right] - 2H\,\mathbf{n}(\mathbf{x}) \cdot (\nabla_{S_I} \cdot \mathbf{M}_S(\mathbf{x}))\mathbf{n}(\mathbf{x}). \tag{4}$$

The unit vector $\mathbf{n}$ in the above equation is normal to S_I, and it is assumed to point away from the inhomogeneity. The square brackets indicate the jump of the field quantities across the interface, defined as their value on the side towards which vector $\mathbf{n}$ points minus their value on the side from which it points; ∇_{S_I} is the surface gradient operator; $2H = \mathrm{tr}\mathbf{B}(\mathbf{x})$ is the main curvature; $\mathbf{B}(\mathbf{x}) = -\nabla_{S_I}\mathbf{n}(\mathbf{x})$ is the curvature tensor; and the surface membrane stress tensor $\boldsymbol{\sigma}_S$ [12,13] is defined as

$$\boldsymbol{\sigma}_S(\mathbf{x}) = \tau_0 \overset{2}{\mathbf{I}}_S + 2[\mu_S - \tau_0]\boldsymbol{\varepsilon}_S(\mathbf{x}) + [\lambda_S + \tau_0]\mathrm{tr}(\boldsymbol{\varepsilon}_S(\mathbf{x}))\overset{2}{\mathbf{I}}_S + \tau_0\nabla_S\mathbf{u}(\mathbf{x}), \tag{5}$$

where $\boldsymbol{\varepsilon}_S$ is the interface/surface membrane strain tensor, $\overset{2}{\mathbf{I}}_S$ represents the second-rank identity tensor in the plane tangent to the surface, τ_0 is the magnitude of the deformation-independent (residual) surface/interfacial tension (assumed "hydrostatic" and constant in Gurtin–Murdoch model), λ_S and μ_S are surface Lamé parameters, while $\nabla_S\mathbf{u}(\mathbf{x})$ denotes the surface gradient of the interface displacement field.

The surface couple stress tensor $\mathbf{M}_S$ (moments), which described surface bending [35,36,40], has the following form:

$$\mathbf{M}_S(\mathbf{x}) = 2\mu_B\,\boldsymbol{\kappa}_S(\mathbf{x}) + \lambda_B\mathrm{tr}(\boldsymbol{\kappa}_S(\mathbf{x}))\overset{2}{\mathbf{I}}_S, \tag{6}$$

The symbols λ_B and μ_B are the material parameters describing the bending stiffness of the (isotropic) material surface. The surface strain tensor $\boldsymbol{\varepsilon}_S$ and the bending strain measure (tensor representing changes of curvature due to bending) $\boldsymbol{\kappa}_S$ are as follows:

$$\boldsymbol{\varepsilon}_S(\mathbf{x}) = \mathrm{sym}\!\left(\overset{2}{\mathbf{I}}_S(\mathbf{x}) \cdot \nabla_S\mathbf{u}(\mathbf{x})\right), \tag{7}$$

$$\boldsymbol{\kappa}_S = \mathrm{sym}\!\left(\overset{2}{\mathbf{I}}_S(\mathbf{x}) \cdot \nabla_S\vartheta(\mathbf{x})\right), \tag{8}$$

in which $\vartheta(\mathbf{x})$ represents rotation of the surface (more specifically, it is displacement of the tip of vector $\mathbf{n}(\mathbf{x})$ due to rotation of the surface)

$$\vartheta(\mathbf{x}) = \nabla_{S_I}(\mathbf{n}(\mathbf{x}) \cdot \mathbf{u}(\mathbf{x})) + \mathbf{B}(\mathbf{x}) \cdot \mathbf{u}(\mathbf{x}). \tag{9}$$

In the case of Steigmann–Ogden interface Equations (3)–(9), the surface energy can be represented as

$$E_{\text{int}} = U_T + U_B, \tag{10}$$

where U_T and U_B are the energies related to surface tension and surface bending:

$$U_T = \frac{1}{2}\oint_{S_I}\left[2\bar{\mu}_S \varepsilon_S : \varepsilon_S + \bar{\lambda}_S \operatorname{tr}(\varepsilon_S)^2 + \tau_0 \nabla_S \mathbf{u} : \nabla_S \mathbf{u}\right]dS, \tag{11}$$

$$U_B = \frac{1}{2}\oint_{S_I}\left[2\mu_B \kappa : \kappa + \lambda_B \left(\operatorname{tr}\kappa\right)^2\right]dS. \tag{12}$$

In [42], it is shown that, for cylindrical approximation of short fibers described by Gurtin–Murdoch surface expressed in Equations (3)–(9), if $\mathbf{M}_S(\mathbf{x}) = 0$, the stiffness tensors $\mathbf{C}_{\text{eq}}$ have transversely isotropic symmetry, characterized by 5 independent constants, and have the following form:

$$\mathbf{C}_{\text{eq}} = \mathbf{C}_1 + \hat{\mathbf{C}}_T, \tag{13}$$

where $\mathbf{C}_1$ is the stiffness tensor of the original inhomogeneity while $\hat{\mathbf{C}}_T$ represents an additional contribution of surface elasticity to the properties of an equivalent cylindrical fiber (see details in [42]). This specific form of Equation (13) results from the fact that the Gurtin–Murdoch model assumes vanishingly thin interphase and preserves kinematic continuity, so in Equation (2), $\varepsilon_1 = \varepsilon_{\text{eq}}$ —a property that is also preserved in the Steigmann–Ogden interface model and is exploited subsequently. The transversely isotropic symmetry of equivalent elasticity stiffness tensor will be subsequently written in Voigt's notation assuming the following identification scheme:

$$11 \rightarrow 1\,,\ 22 \rightarrow 2\,,\ 33 \rightarrow 3\,,\ 23,32 \rightarrow 4\,,\ 13,31 \rightarrow 5\,,\ 12,21 \rightarrow 6, \tag{14}$$

with index 3 denoting the longitudinal axis of the fiber. As shown in [42], the expressions for the six non-vanishing components of matrix $\hat{\mathbf{C}}_T$, five of which are independent, have the following forms:

$$\hat{C}_{T[11]} = \hat{C}_{T[22]} = \left(\frac{3}{4r} + \frac{2}{l}\right)\left[2\bar{\mu}_S + \bar{\lambda}_S\right],\ \hat{C}_{T[33]} = \frac{2\left[2\bar{\mu}_S + \bar{\lambda}_S\right]}{r},$$

$$\hat{C}_{T[12]} = \frac{\left[2\bar{\mu}_S + \bar{\lambda}_S\right]}{4r} + \frac{2\bar{\lambda}_S}{l},\ \hat{C}_{T[13]} = \hat{C}_{T[23]} = \frac{\bar{\lambda}_S}{r},$$

$$\hat{C}_{T[44]} = \hat{C}_{T[55]} = \frac{\bar{\mu}_S}{r},\ \hat{C}_{T[66]} = \frac{1}{2}\left[\hat{C}_{T[11]} - \hat{C}_{T[12]}\right] = \frac{\left[2\bar{\mu}_S + \bar{\lambda}_S\right]}{4r} + \frac{2\bar{\mu}_S}{l}, \tag{15}$$

where $\bar{\lambda}_S = \lambda_S + \tau_0$ and $\bar{\mu}_S = \mu_S - \tau_0$ are the modified Lamé parameters of isotropic tensor of surface elasticity, appearing as a result of the surface tension contribution in Equations (5) and (11); r is the radius of the cylinder; and l is its length (see Figure 1). In the adopted Steigmann–Ogden interface model, the tenor $\mathbf{C}_{\text{eq}}$ should be also transversely isotropic and can be defined as

$$\mathbf{C}_{\text{eq}} = \mathbf{C}_1 + \hat{\mathbf{C}}_T + \hat{\mathbf{C}}_B, \tag{16}$$

where $\hat{\mathbf{C}}_B$ is a contribution of surface bending.

The development neglecting the term $\mathbf{M}_S$ of Equations (4) and (6) was presented in [42]. Here, the focus of evaluation of the properties of EEI is on the contribution of surface bending. The use of the complete Equation (4) in analysis may turn out to be important in some practical applications, where bending of the surface should be accounted for.

Inclusion of the complete Equation (4) within the framework of the EEI is outlined in the next subsection, with some supporting derivations presented in the related Appendix A.

2.3. Contribution of the Surface Bending to the Energy of Equivalent Cylinder

2.3.1. Evaluation of the Surface Energy Related to the Bending

In order to determine the surface contribution described by Equation (12), the tensor of curvature changes will be evaluated first.

It is assumed that the strains, ε_{eq}, that an inhomogeneity is subjected to are constant. Under those conditions, displacements in the surface of that inhomogeneity can be expressed as

$$\mathbf{u}(\xi^\Lambda) = \varepsilon_{eq} \cdot \mathbf{r}(\xi^\Lambda). \tag{17}$$

where $\mathbf{r}(\xi^\Lambda)$ is the position vector of a point on that surface which is locally parameterized by ξ^Λ, $\Lambda \in \{1, 2\}$. Consequently, cf. [44]

$$\nabla_S \mathbf{u} = (\varepsilon_{eq} \cdot \mathbf{r})_{,\Lambda} \otimes \mathbf{G}^\Lambda = \varepsilon_{eq} \cdot (\mathbf{r}_{,\Lambda} \otimes \mathbf{G}^\Lambda) = \varepsilon_{eq} \cdot (\mathbf{G}_\Lambda \otimes \mathbf{G}^\Lambda) = \varepsilon_{eq} \cdot \overset{2}{\mathbf{I}}_S. \tag{18}$$

where $\mathbf{G}_\Lambda = \mathbf{r}_{,\Lambda}$ are the vectors of the natural basis associated with the parametrization ξ^Λ (tangent to the surface) and $\mathbf{G}^\Lambda$ is the vectors of the dual, or reciprocal, basis also tangent to the surface) satisfying the condition $\mathbf{G}_\Lambda \cdot \mathbf{G}^\Lambda = \delta_\Lambda{}^\Lambda$, with $\delta_\Lambda{}^\Lambda$ being the "Kronecker delta".

The tensor of curvature changes is determined as

$$\kappa = \mathrm{sym}\!\left(\overset{2}{\mathbf{I}}_S \cdot \nabla_S \vartheta\right), \tag{19}$$

with

$$\vartheta = \omega_N \cdot \mathbf{n}, \tag{20}$$

where

$$\omega_N = -(\mathbf{n} \cdot \nabla_S \mathbf{u}) \otimes \mathbf{n}. \tag{21}$$

Considering Equation (18), ϑ can be defined as

$$\vartheta = -\mathbf{n} \cdot \varepsilon_{eq} \cdot \overset{2}{\mathbf{I}}_S = -\varepsilon_{eq} : \mathbf{n} \otimes \overset{2}{\mathbf{I}}_S, \tag{22}$$

which gives

$$\nabla_S \vartheta = -\varepsilon_{eq} : \nabla_S\!\left(\mathbf{n} \otimes \overset{2}{\mathbf{I}}_S\right); \tag{23}$$

$$\kappa = -\mathrm{sym}\!\left[\overset{2}{\mathbf{I}}_S \cdot \left(\varepsilon_{eq} : \nabla_S\!\left(\mathbf{n} \otimes \overset{2}{\mathbf{I}}_S\right)\right)\right]. \tag{24}$$

The above formula indicates that ε_{eq} contracted with the first 2 vectors of $\nabla_S(\mathbf{n} \otimes \overset{2}{\mathbf{I}}_S)$ and $\overset{2}{\mathbf{I}}_S$ operate on the third vector of dyadic product in $\nabla_S(\mathbf{n} \otimes \overset{2}{\mathbf{I}}_S)$. This means that multiplication by $\overset{2}{\mathbf{I}}_S$ eliminates $\mathbf{n} \otimes \mathbf{G}_\Lambda \otimes \mathbf{n} \otimes \mathbf{G}^\Lambda$ ($\mathbf{G}_\Lambda \perp \mathbf{n}$) and the remaining two parts are unchanged. Therefore,

$$\boldsymbol{\kappa} = -\mathrm{sym}\!\left[\mathbf{I}_S \cdot (\nabla_S(\mathbf{n} \otimes \mathbf{I}_S))^T : \boldsymbol{\varepsilon}_{eq}\right]$$

$$= -\mathrm{sym}\!\left[\left(-B_\Lambda{}^\Pi \mathbf{G}^\Lambda \otimes \mathbf{G}^\Delta \otimes \mathbf{G}_\Pi \otimes \mathbf{G}_\Lambda + B_{\Lambda\Lambda}\mathbf{G}^\Lambda \otimes \mathbf{G}^\Delta \otimes \mathbf{n} \otimes \mathbf{n}\right) : \boldsymbol{\varepsilon}_{eq}\right]. \tag{25}$$

Evaluation of the components of the tensors $B_{\Lambda\Lambda}$ and $B_\Lambda{}^\Pi$ is illustrated in Appendix A, where curvature tensors for spherical inhomogeneity of radius r are given in Table A1. Considering values $B_{\Lambda\Lambda}$ and $B_\Lambda{}^\Delta$ from Table A1, the tensor of curvature changes can be defined as

$$\boldsymbol{\kappa} = -\frac{1}{r}\big[(\overline{\mathbf{G}}_1 \otimes \overline{\mathbf{G}}_1 \otimes \overline{\mathbf{G}}_1 \otimes \overline{\mathbf{G}}_1 + \frac{1}{2}(\overline{\mathbf{G}}_2 \otimes \overline{\mathbf{G}}_1 + \overline{\mathbf{G}}_1 \otimes \overline{\mathbf{G}}_2) \otimes \overline{\mathbf{G}}_1 \otimes \overline{\mathbf{G}}_2 -$$

$$-\,\overline{\mathbf{G}}_1 \otimes \overline{\mathbf{G}}_1 \otimes \mathbf{n} \otimes \mathbf{n}) : \boldsymbol{\varepsilon}_{eq}\big]. \tag{26}$$

Then, the surface strain energy in the case of the Steigmann–Ogden model of interface is defined as (see, for details, [35,36,40])

$$E_S = \frac{1}{2}\oint_{S_I}\left[2\overline{\mu}_S \boldsymbol{\varepsilon}_S : \boldsymbol{\varepsilon}_S + \overline{\lambda}_S \,\mathrm{tr}(\boldsymbol{\varepsilon}_S)^2 + \tau_0 \nabla_S \mathbf{u} : \nabla_S \mathbf{u} + 2\mu_B \boldsymbol{\kappa} : \boldsymbol{\kappa} + \lambda_B\,(\mathrm{tr}\,\boldsymbol{\kappa})^2\right]\!dS. \tag{27}$$

The first three terms of the above integrand are identical to the surface strain energy given by the Gurtin–Murdoch model, and properties of equivalent inhomogeneities related to these terms are determined in [42]. The last two terms of Equation (27) represent the surface strain energy related to surface bending. In the next section, the working formula for the properties of the equivalent inhomogeneity is presented.

2.3.2. Constitutive Tensor of the Energy-Equivalent Cylinder

Considering Equation (26), the last two terms of Equation (27) are given as

$$\boldsymbol{\kappa} : \boldsymbol{\kappa} = \frac{1}{r^2}\boldsymbol{\varepsilon}_{eq} : \big[\overline{\mathbf{G}}_1 \otimes \overline{\mathbf{G}}_1 \otimes \overline{\mathbf{G}}_1 \otimes \overline{\mathbf{G}}_1 + \overline{\mathbf{G}}_1 \otimes \overline{\mathbf{G}}_2 \otimes \overline{\mathbf{G}}_1 \otimes \overline{\mathbf{G}}_2 - \overline{\mathbf{G}}_1 \otimes \overline{\mathbf{G}}_1 \otimes \mathbf{n} \otimes \mathbf{n} -$$

$$-\,\mathbf{n} \otimes \mathbf{n} \otimes \overline{\mathbf{G}}_1 \otimes \overline{\mathbf{G}}_1 + \mathbf{n} \otimes \mathbf{n} \otimes \mathbf{n} \otimes \mathbf{n}\big] : \boldsymbol{\varepsilon}_{eq}; \tag{28}$$

$$(\mathrm{tr}\,\boldsymbol{\kappa})^2 = \frac{1}{r^2}\boldsymbol{\varepsilon}_{eq} : \big[\overline{\mathbf{G}}_1 \otimes \overline{\mathbf{G}}_1 \otimes \overline{\mathbf{G}}_1 \otimes \overline{\mathbf{G}}_1 - 2\overline{\mathbf{G}}_1 \otimes \overline{\mathbf{G}}_1 \otimes \mathbf{n} \otimes \mathbf{n} + \mathbf{n} \otimes \mathbf{n} \otimes \mathbf{n} \otimes \mathbf{n}\big] : \boldsymbol{\varepsilon}_{eq}. \tag{29}$$

The surface energy of Equation (27) is a sum of the surface tension and the surface bending Equations (11) and (12), and we focus only on the latter:

$$U_B = \frac{1}{2}\oint_S\left[2\mu_B \boldsymbol{\kappa} : \boldsymbol{\kappa} + \lambda_B\,(\mathrm{tr}\,\boldsymbol{\kappa})^2\right]\!dS = U_{\mu_B} + U_{\lambda_B}, \tag{30}$$

where $U_{\mu_B} = \frac{1}{2}\oint_S[2\mu_B \boldsymbol{\kappa} : \boldsymbol{\kappa}]dS$ is defined as

$$U_{\mu_B} = \frac{2\mu_B}{2r^2}\oint_{S_I}\Big[\boldsymbol{\varepsilon}_{eq} : (\overline{\mathbf{G}}_1 \otimes \overline{\mathbf{G}}_1 \otimes \overline{\mathbf{G}}_1 \otimes \overline{\mathbf{G}}_1 + \overline{\mathbf{G}}_1 \otimes \overline{\mathbf{G}}_2 \otimes \overline{\mathbf{G}}_1 \otimes \overline{\mathbf{G}}_2 - \overline{\mathbf{G}}_1 \otimes \overline{\mathbf{G}}_1 \otimes \mathbf{n} \otimes \mathbf{n} -$$

$$-\,\mathbf{n} \otimes \mathbf{n} \otimes \overline{\mathbf{G}}_1 \otimes \overline{\mathbf{G}}_1 + \mathbf{n} \otimes \mathbf{n} \otimes \mathbf{n} \otimes \mathbf{n}) : \boldsymbol{\varepsilon}_{eq}\Big]dS. \tag{31}$$

and $U_{\lambda_B} = \frac{1}{2} \oint_S \left[\lambda_B \, (\text{tr } \kappa)^2 \right] dS$ is

$$U_{\lambda_B} = \frac{\lambda_B}{2r^2} \oint_{S_I} \left[\varepsilon_{\text{eq}} : (\overline{\mathbf{G}}_1 \otimes \overline{\mathbf{G}}_1 \otimes \overline{\mathbf{G}}_1 \otimes \overline{\mathbf{G}}_1 - 2\overline{\mathbf{G}}_1 \otimes \overline{\mathbf{G}}_1 \otimes \mathbf{n} \otimes \mathbf{n} + \mathbf{n} \otimes \mathbf{n} \otimes \mathbf{n} \otimes \mathbf{n}) : \varepsilon_{\text{eq}} \right] dS. \tag{32}$$

These last formulas in Equations (30)–(32) can be put in Equation (10), and the following form of the surface energy yields

$$E_{\text{int}} = U_T + U_{\mu_B} + U_{\lambda_B} = \varepsilon_{\text{eq}} : (\mathbf{K}_T + \mathbf{K}_{\mu_B} + \mathbf{K}_{\lambda_B}) : \varepsilon_{\text{eq}}. \tag{33}$$

where

$$U_T = \varepsilon_{\text{eq}} : \mathbf{K}_T : \varepsilon_{\text{eq}}; \quad U_{\mu_B} = \varepsilon_{\text{eq}} : \mathbf{K}_{\mu_B} : \varepsilon_{\text{eq}}; \quad U_{\lambda_B} = \varepsilon_{\text{eq}} : \mathbf{K}_{\lambda_B} : \varepsilon_{\text{eq}}. \tag{34}$$

The last result for E_{int} and the energy equivalence expressed by Equation (2) leads to the following formula for the effective moduli of equivalent inhomogeneity:

$$\mathbf{C}_{\text{eq}} = \mathbf{C}_1 + \frac{1}{V_I}(\mathbf{K}_T + \mathbf{K}_{\mu_B} + \mathbf{K}_{\lambda_B}) = \mathbf{C}_1 + \hat{\mathbf{C}}_T + \hat{\mathbf{C}}_B. \tag{35}$$

Then, the problem of properties of equivalent inhomogeneities is reduced to evaluation of the components of the above tensors $\mathbf{K}_{\mu_B}$ and $\mathbf{K}_{\lambda_B}$, which is illustrated in Appendix B Equations (A35)–(A37). Contribution of surface bending to the stiffness tensor Equation (35) in this case is

$$\hat{C}_{B[11]} = \hat{C}_{B[22]} = \frac{\lambda_B + 2\mu_B}{r^3}, \; \hat{C}_{B[33]} = 0,$$

$$\hat{C}_{B[12]} = -\frac{\lambda_B + 2\mu_B}{r^3}, \; \hat{C}_{B[13]} = \hat{C}_{B[23]} = 0,$$

$$\hat{C}_{B[44]} = \hat{C}_{B[55]} = \frac{\mu_B}{r^3}, \; \hat{C}_{B[66]} = \frac{1}{2}\left[\hat{C}_{B[11]} - \hat{C}_{B[12]}\right] = \frac{\lambda_B + 2\mu_B}{r^3}, \tag{36}$$

where λ_B and μ_B are additional material parameters describing the bending stiffness of the material surface in Equation (6). In the presence of $\mathbf{M}_S$ in Equation (4), the tenor $\mathbf{C}_{\text{eq}}$ is also transversely isotropic and its constants are defined in Equation (35), where $\hat{\mathbf{C}}_T$ is defined in Equation (15).

Remark 1. *It should be noted that the properties of equivalent cylindrical fibers can be used in combination with any homogenization method developed without accounting for interfaces.*

3. Comparison with the Existing Results for the Cylinder of Infinite Length with Gurtin–Murdoch and Steigmann–Ogden Interfaces

To validate the proposed approach, the equivalent properties of cylinder of infinite length are obtained as a limiting case and compared with two-dimensional solutions of the problem, which are the only currently available results for cylindrical inhomogeneities with Gurtin–Murdoch and Steigmann–Ogden surfaces.

In the limit $l \to \infty$ and $\lambda_B = \mu_B = 0$, one obtains the results for an equivalent infinite cylindrical fiber with Gurtin–Murdoch interface. The independent constants of matrix $\hat{\mathbf{C}}_T$ in this case are

$$\hat{C}_{T[11]} = \hat{C}_{T[22]} = \frac{3}{4r}\left[2\overline{\mu}_S + \overline{\lambda}_S\right], \; \hat{C}_{T[33]} = \frac{2\left[2\overline{\mu}_S + \overline{\lambda}_S\right]}{r},$$

$$\hat{C}_{T[12]} = \frac{\left[2\overline{\mu}_S + \overline{\lambda}_S\right]}{4r}, \; \hat{C}_{T[13]} = \hat{C}_{T[23]} = \frac{\overline{\lambda}_S}{r},$$

$$\hat{C}_{T[44]} = \hat{C}_{T[55]} = \frac{\overline{\mu}_S}{r}, \ \hat{C}_{T[66]} = \frac{1}{2}\left[\hat{C}_{T[11]} - \hat{C}_{T[12]}\right] = \frac{\left[2\overline{\mu}_S + \overline{\lambda}_S\right]}{4r}, \tag{37}$$

Four out of the above five constants, $\hat{C}_{T[11]}, \hat{C}_{T[33]}, \hat{C}_{T[13]}, \hat{C}_{T[44]}$, can be presented in the form of Hill's phase moduli [45], and in that form, they are exactly the same as those presented in [15,16]. In those publications, the properties of an equivalent infinite cylindrical fiber with Gurtin–Murdoch interface have been determined first using the concept of neutral inhomogeneity [46]. The authors subsequently observed that the same equivalent properties are obtained if the constitutive tensor of the fibers is augmented by the terms shown in Equations (37). Such an agreement with the values obtained "a posteriori" and by a different approach furnishes an additional support for the concept of equivalent inhomogeneity presented herein.

The fifth constant, the transverse shear modulus, constitutes an exception in the sense that, in [15], it could not be determined by the same approach as the other four, i.e., by a neutral composite cylinder approach or composite cylinder assembly [5,16]. Thus, the generalized self-consistent method [47] has been employed in [16] instead, which turned out not to allow for identification of the contribution of surface elasticity to the fifth constant of the equivalent cylinder.

The properties of an equivalent infinite cylindrical fiber with a Steigmann–Ogden interface can be also compared with the two-dimensional solutions presented in [36]. Two of the constants listed in Equations (36) and (37), $\hat{C}_{B[11]} + \hat{C}_{B[12]}$ and $\hat{C}_{B[66]}$, $\hat{C}_{T[11]} + \hat{C}_{T[12]}$ and $\hat{C}_{T[66]}$ for surface bending and surface tension, were presented in the form of the plane bulk modulus and transverse shear modulus (Hill's notation) and compared with those obtained in [36] for two-dimensional solutions of the problem (limit if $l \rightarrow \infty$). Properties of equivalent circular inhomogeneity [36] are obtained based on well-known elasticity solutions for two complementary problems: one of the circular discs subjected to the unknown tractions at any boundary point and another one from an infinite matrix subjected to the uniform far-field load and containing a circular hole under the action of unknown boundary tractions. The solutions for both problems can be obtained by the complex variables approach. It is shown in [36] that surface bending does not affect the plane bulk modulus $\hat{C}_{B[11]} + \hat{C}_{B[12]}$, and at the same time, the contribution of the surface tension to the plane bulk modulus of equivalent cylinder presented herein is identical to the results in [36], really, Equation (54) in [46] (in notations adopted in the present article):

$$C_{eq[11]} + C_{eq[12]} = C_{1[11]} + C_{1[12]} + \frac{1}{2r}\left[2\overline{\mu}_S + \overline{\lambda}_S\right]. \tag{38}$$

Comparing Equations (35)–(38), it is evident that

$$\hat{C}_{T[11]} + \hat{C}_{T[12]} = \frac{1}{2r}\left[2\overline{\mu}_S + \overline{\lambda}_S\right].$$

$$\hat{C}_{B[11]} + \hat{C}_{B[12]} = 0, \tag{39}$$

The above Equations (39) are identical to Equations (36) and (37).

Given that $(\lambda_B + 2\mu_B)/r^3$ and $(\overline{\lambda}_S + 2\overline{\mu}_S)/r$ are considerably smaller then Lamé parameters of bulk material (see, e.g., [24,25,36,40]), it is possible to check that first-order approximation of the results for transverse shear moduli $\hat{C}_{B[66]}$ and $\hat{C}_{T[66]}$ presented in [36] (in Equation (59)) coincides with the ones presented here. Thus, Equation (59) in [36] (in notations adopted in the present article) is

$$C_{eq[66]} = C_{1[66]} + \frac{\left[C_{2[66]}(C_{1[11]} + C_{1[12]}) + C_{1[66]}\right](\eta + \gamma) + 12(C_{1[11]} + C_{1[12]})\eta\gamma}{C_{2[66]}(C_{1[11]} + C_{1[12]}) + C_{1[66]} + 12(C_{1[11]} + 3C_{1[12]})(\eta + \gamma)}, \tag{40}$$

where $\eta = (\overline{\lambda}_S + 2\overline{\mu}_S)/r$ and $\gamma = (\lambda_B + 2\mu_B)/r^3$. It should be noted that the properties of an equivalent infinite cylinder are obtained on the basis of a solution for an infinite matrix subjected to the uniform far-field load and containing a circular hole under the action of unknown boundary tractions, and as a result, they depend on the properties of the infinite matrix $C_{2[66]}$. The first-order approximation of Equation (40) is as follows:

$$C_{\text{eq}[66]} = C_{1[66]} + \eta + \gamma, \tag{41}$$

and it is identical to Equations (35)–(37).

4. Conclusions

A mathematical model employing the concept of the EEI [19,41–43] has been generalized to introduce the surface effects described by the Steigmann–Ogden model [22,23] derived within the strain gradient elasticity [35]. A particular focus was centered on accounting for surface bending contribution in the definition of the EEI.

The properties of an equivalent cylinder of finite length with the Steigmann–Ogden model of interface is determined based on the corresponding definition of surface energy, which includes both surface tension and surface bending. As typically done in Hill's equivalence principle, a uniform state of strains within the cylinder is assumed. The stiffness tensor of the equivalent cylinder has transversely isotropic symmetry, and five independent constants of this tensor are presented in a closed form.

Unfortunately, due to a lack of solutions for problems involving finite-length cylindrical fibers, the main results presented herein could not be verified by direct comparisons. This could be accomplished only by the asymptotic transformation of those results to obtain equivalent stiffness tensors for infinite cylindrical inhomogeneity. As shown in Section 3, in the asymptotic limit, the results obtained in this work are in a good agreement with those of [15,16] obtained for infinite cylindrical inhomogeneities with Gurtin–Murdoch interface (by solving a number of two-dimensional problems). They are also in good agreement with results obtained for the plane bulk and transverse shear moduli derived for the two-dimensional problem of circular inhomogeneities with the Steigmann–Ogden model of interface.

To conclude, it is worth mentioning that the definition of the EEI is general and can be used in the case of inhomogeneities of shapes other than cylindrical, e.g., ellipsoidal. It can be very naturally combined with any homogenization method developed for composite materials without accounting for interface and appears to be potentially amenable for inclusion of models other than the Gurtin–Murdoch or Steigmann–Ogden interface models. The important characteristic of the proposed approach is its ability to provide closed-form expressions for the properties of equivalent inhomogeneities. Closed-form results are important, especially if the influence of different problem parameters needs to be analyzed.

Author Contributions: The conception of the idea as well as drafting and revising various versions of the manuscript was a truly collaborative effort. The idea was born during discussion of previous jointly coauthored articles. The preparation of the manuscript involved several iterations in which all authors took consecutive turns in reading the manuscript and inserting their comments. The individual authors have been largely responsible for the following technical aspects of the paper: H.S., governing equations of the problem and evaluation of the formula for energy of the system (original inhomogeneity and interphase); L.N., development of equations for determination of the properties of equivalent inhomogeneity and their implementation, and the equivalent properties of cylinders with the Steigmann–Ogden model of interface; H.A., addition of some important references, analysis and interpretation of obtained results, and final approval of the version to be published. All authors have read and agreed to the published version of the manuscript.

Funding: This research was funded by the German Research Foundation (DFG), grant number AL 341/51-1.

Acknowledgments: L.N. and H.A. gratefully acknowledge the financial support by the German Research Foundation (DFG) via project AL 341/51-1.

Conflicts of Interest: The authors declare no conflict of interest.

Appendix A. Components of Curvature Tensors for Cylinder

Let us assume that the surface of interest is locally parameterized by ξ^Λ, $\Lambda \in \{1, 2\}$, and that the position vector of a point on that surface is expressed as $\mathbf{r}(\xi^\Lambda)$. Then, one can define a couple of vectors $\mathbf{G}_\Lambda$:

$$\mathbf{G}_\Lambda = \frac{\partial \mathbf{r}}{\partial \xi^\Lambda} \equiv \mathbf{r}_{,\Lambda}, \tag{A1}$$

which forms the vector basis in the linear space tangent to the surface S, called the natural basis. Another basis in the same tangent space, denoted by $\mathbf{G}^\Delta$ and called dual or reciprocal, is defined as follows:

$$\mathbf{G}_\Lambda \cdot \mathbf{G}^\Delta = \delta^\Delta{}_\Lambda, \tag{A2}$$

where $\cdot$ represents the "dot" (or "inner") product of vectors and $\delta^\Delta{}_\Lambda$ is the Kronecker "delta". The bases $\mathbf{G}_\Lambda$ and $\mathbf{G}^\Delta$ are functions of ξ^Λ, and their derivatives can be expressed by the well-known Gauss–Weingarten formulas (see [44], for example). For the natural basis, these formulas are (cf. Equation (A1) for notations)

$$\mathbf{G}_{\Lambda,\Sigma} = \Gamma_{\Lambda\Sigma}{}^\Delta \mathbf{G}_\Delta + B_{\Lambda\Sigma}\mathbf{n} \equiv \Gamma_{\Lambda\Sigma}{}^1 \mathbf{G}_1 + \Gamma_{\Lambda\Sigma}{}^2 \mathbf{G}_2 + B_{\Lambda\Sigma}\mathbf{n}, \tag{A3}$$

with a unit vector $\mathbf{n}$ normal to the surface. Here (as shown in the above equation), an index repeated in the subscript and superscript position implies summation, $\Gamma_{\Lambda\Sigma}{}^\Delta = \mathbf{G}_{\Lambda,\Sigma} \cdot \mathbf{G}^\Delta$ are the so-called Christoffel symbols (of the second kind), and the components of the local curvature tensor are

$$B_{\Lambda\Sigma} = \mathbf{G}_{\Lambda,\Sigma} \cdot \mathbf{n}. \tag{A4}$$

Equation (A3) together with Equation (A1) imply that $B_{\Lambda\Sigma} = B_{\Sigma\Lambda}$, whereas definition of the Christoffel symbols and Equation (A1) imply the following symmetry property: $\Gamma_{\Lambda\Sigma}{}^\Omega = \Gamma_{\Sigma\Lambda}{}^\Omega$. The analogical formulas for the derivatives of vectors of the dual basis are

$$\mathbf{G}^\Delta{}_{,\Sigma} = -\Gamma_{\Lambda\Sigma}{}^\Delta \mathbf{G}^\Lambda - B^\Delta{}_\Sigma \mathbf{n} \equiv -\Gamma_{1\Sigma}{}^\Delta \mathbf{G}^1 - \Gamma_{2\Sigma}{}^\Delta \mathbf{G}^2 - B^\Delta{}_\Sigma \mathbf{n}, \tag{A5}$$

where $B^\Delta{}_\Sigma$ is the so-called mixed components of the local curvature tensor. The curvature tensor $\mathbf{B}$ can be represented as

$$\mathbf{B} = B_{\Lambda\Lambda}\mathbf{G}^\Delta \otimes \mathbf{G}^\Lambda = B^{\Delta\Lambda}\mathbf{G}_\Delta \otimes \mathbf{G}_\Lambda = B_\Delta{}^\Lambda \mathbf{G}^\Delta \otimes \mathbf{G}_\Lambda = B^\Delta{}_\Lambda \mathbf{G}_\Delta \otimes \mathbf{G}^\Lambda. \tag{A6}$$

In the above equation, double summation is implied and the (indexed) coefficients multiplying the dyadic are various components of tensor $\mathbf{B}$. They all can be different, but they are related to each other by transformation formulas involving the so-called gram matrices related to the natural or dual bases. Those matrices are defined as follows:

$$G_{\Delta\Lambda} = \mathbf{G}_\Delta \cdot \mathbf{G}_\Lambda, \ G^{\Delta\Lambda} = \mathbf{G}^\Delta \cdot \mathbf{G}^\Lambda. \tag{A7}$$

Using the relationship between various components of the curvature tensor $\mathbf{B}$, one can present the following:

$$B_\Delta{}^\Lambda = B_{\Delta\Sigma}G^{\Sigma\Lambda}. \tag{A8}$$

In the case of cylindrical inhomogeneity of radius r, the position vector $\mathbf{R}$ of a point on the surface of the inhomogeneity and the corresponding unit vector $\mathbf{n}$, normal to that surface may be expressed in cylindrical coordinates as follows:

$$\mathbf{R} = \begin{bmatrix} r\cos\varphi \\ r\sin\varphi \\ z \end{bmatrix}, \ 0 \leq \varphi \leq 2\pi. \tag{A9}$$

Consequently, the local vectors of the natural basis $\mathbf{G}_\Lambda$ are

$$\mathbf{G}_1 = \mathbf{R}_{,\varphi} = r\begin{bmatrix} -\sin\varphi \\ \cos\varphi \\ 0 \end{bmatrix}; \ \mathbf{G}_2 = \mathbf{R}_{,z} = \begin{bmatrix} 0 \\ 0 \\ 1 \end{bmatrix}; \ \mathbf{G}_3 = \mathbf{R}_{,r} = \mathbf{n} = \begin{bmatrix} \cos\varphi \\ \sin\varphi \\ 0 \end{bmatrix}, \tag{A10}$$

and

$$\mathbf{G}_{1,1} = \mathbf{R}_{,\varphi\varphi} = r\begin{bmatrix} -\cos\varphi \\ -\sin\varphi \\ 0 \end{bmatrix}; \ \mathbf{G}_{2,1} = \mathbf{R}_{,z\varphi} = \begin{bmatrix} 0 \\ 0 \\ 0 \end{bmatrix}; \ \mathbf{G}_{2,2} = \mathbf{R}_{,zz} = \begin{bmatrix} 0 \\ 0 \\ 0 \end{bmatrix}. \tag{A11}$$

The dual basis is defined as

$$\mathbf{G}^1 = \frac{1}{r}\begin{bmatrix} -\sin\varphi \\ \cos\varphi \\ 0 \end{bmatrix}, \ \mathbf{G}^2 = \begin{bmatrix} 0 \\ 0 \\ 1 \end{bmatrix}, \ \mathbf{G}_3 = \mathbf{G}^3 = \mathbf{n} = \begin{bmatrix} \cos\varphi \\ \sin\varphi \\ 0 \end{bmatrix}. \tag{A12}$$

Then, the curvature tensors for cylindrical inhomogeneity of radius r are given:

Table A1. Curvature tensors for a cylinder of radius r.

Λ	Δ	$B_{\Lambda\Delta}$	$B_\Lambda{}^\Delta$
1	1	$-r$	$-1/r$
1	2	0	0
2	2	0	0
2	1	0	0

The local vectors of the natural basis $\mathbf{G}_\Lambda$ on the circular part of the cylinder's surface (i.e., on its two ends) are

$$\mathbf{G}_1 = \mathbf{R}_{,r} = \begin{bmatrix} \cos\varphi \\ \sin\varphi \\ 0 \end{bmatrix}; \ \mathbf{G}_2 = \mathbf{R}_{,\varphi} = r\begin{bmatrix} -\sin\varphi \\ \cos\varphi \\ 0 \end{bmatrix}; \ \mathbf{G}_3 = \mathbf{R}_{,z} = \mathbf{n} = \begin{bmatrix} 0 \\ 0 \\ 1 \end{bmatrix}, \tag{A13}$$

and

$$\mathbf{G}_{1,1} = \mathbf{R}_{,rr} = \begin{bmatrix} 0 \\ 0 \\ 0 \end{bmatrix}; \ \mathbf{G}_2 = \mathbf{R}_{,\varphi r} = \begin{bmatrix} -\sin\varphi \\ \cos\varphi \\ 0 \end{bmatrix}; \ \mathbf{G}_2 = \mathbf{R}_{,\varphi\phi} = r\begin{bmatrix} -\cos\varphi \\ -\sin\varphi \\ 0 \end{bmatrix}. \tag{A14}$$

Substituting (A14) in (A4), it is seen that the curvature tensor of the circular part of the cylinder's surface (two ends of cylinder) is as follows:

$$B_{\Lambda\Sigma} = 0. \tag{A15}$$

As a result, we have that the contribution of the circular part of the cylinder's surface to surface bending of cylindrical inhomogeneity is equal to zero.

Appendix B. Properties of the Energy-Equivalent Cylinder of Finite Length Accounting for Surface Bending

For illustration of some technical details, $\mathbf{K}_{\mu_B}$ and $\mathbf{K}_{\lambda_B}$ of Equations (31)–(34) are evaluated in this Appendix. In addition to $\mathbf{K}_{\mu_B}$ and $\mathbf{K}_{\lambda_B}$, the contribution of surface tension to properties of equivalent inhomogeneity includes another term $\mathbf{K}_T$ present in Equation (34); however, evaluation of this term is presented in [42].

Assuming that inhomogeneities are cylinders of radius r and length l (Figure 1) and using the cylindrical coordinate system in Equation (A9), $\mathbf{K}_{\mu_B}$ of Equations (31) and (33) is described by the following:

$$K_{\mu_B} = \frac{2\mu_B}{\pi r^3 l} \int_0^l \int_0^{2\pi} [\overline{\mathbf{G}}_1 \otimes \overline{\mathbf{G}}_1 \otimes \overline{\mathbf{G}}_1 \otimes \overline{\mathbf{G}}_1 + \overline{\mathbf{G}}_1 \otimes \overline{\mathbf{G}}_2 \otimes \overline{\mathbf{G}}_1 \otimes \overline{\mathbf{G}}_2 - \overline{\mathbf{G}}_1 \otimes \overline{\mathbf{G}}_1 \otimes \mathbf{n} \otimes \mathbf{n} -$$

$$- \mathbf{n} \otimes \mathbf{n} \otimes \overline{\mathbf{G}}_1 \otimes \overline{\mathbf{G}}_1 + \mathbf{n} \otimes \mathbf{n} \otimes \mathbf{n} \otimes \mathbf{n}] d\varphi\, dz; \tag{A16}$$

$$K_{\mu_B[1111]} = K_{\mu_B[2222]} = \frac{2\mu_B}{\pi r^3 l} \int_0^l \int_0^{2\pi} [\sin\varphi^4 - 2\sin\varphi^2 \cos\varphi^2 + \cos\varphi^4] d\varphi\, dz = \frac{2\mu_B}{r^3}; \tag{A17}$$

$$K_{\mu_B[3333]} = \frac{2\mu_B}{\pi r^2 l} \int_0^l \int_0^{2\pi} [0] d\varphi\, dz; \tag{A18}$$

$$K_{\mu_B[1122]} = \frac{2\mu_B}{\pi r^3 l} \int_0^l \int_0^{2\pi} [2\sin\varphi^2 \cos\varphi^2 - \sin\varphi^4 - \cos\varphi^4] d\varphi\, dz = -\frac{2\mu_B}{r^3}; \tag{A19}$$

$$K_{\mu_B[1133]} = K_{\mu_B[2233]} = \frac{2\mu_B}{\pi r^3 l} \int_0^l \int_0^{2\pi} [0] d\varphi\, dz = 0; \tag{A20}$$

$$K_{\mu_B[1212]} = \frac{2\mu_B}{\pi r^3 l} \int_0^l \int_0^{2\pi} [4\sin\varphi^2 \cos\varphi^2] d\varphi\, dz = \frac{2\mu_B}{r^3}; \tag{A21}$$

$$K_{\mu_B[2112]} = \frac{2\mu_B}{\pi r^3 l} \int_0^l \int_0^{2\pi} [4\sin\varphi^2 \cos\varphi^2] d\varphi\, dz = \frac{2\mu_B}{r^3}; \tag{A22}$$

$$K_{\mu_B[1313]} = K_{\mu_B[2323]} = \frac{2\mu_B}{\pi r^3 l} \int_0^l \int_0^{2\pi} [\sin\varphi^2] d\varphi\, dz = \frac{2\mu_B}{r^3}; \tag{A23}$$

$$K_{\mu_B[3113]} = K_{\mu_B[3223]} = \frac{2\mu_B}{\pi r^3 l} \int_0^l \int_0^{2\pi} [0] d\varphi\, dz = 0. \tag{A24}$$

It is seen that tensor $\mathbf{K}_{\mu_B}$ has transversely isotropic symmetry. $\mathbf{K}_{\lambda_B}$ of Equations (32) and (34) is described by the following:

$$K_{\lambda_B} = \frac{\lambda_B}{\pi r^3 l} \int_0^\pi \int_0^{2\pi} [[\overline{\mathbf{G}}_1 \otimes \overline{\mathbf{G}}_1 \otimes \overline{\mathbf{G}}_1 \otimes \overline{\mathbf{G}}_1 - 2\overline{\mathbf{G}}_1 \otimes \overline{\mathbf{G}}_1 \otimes \mathbf{n} \otimes \mathbf{n} + \mathbf{n} \otimes \mathbf{n} \otimes \mathbf{n} \otimes \mathbf{n}] d\varphi dz; \tag{A25}$$

$$K_{\lambda_B[1111]} = K_{\lambda_B[2222]} = \frac{\lambda_B}{\pi r^3 l} \int_0^l \int_0^{2\pi} [\sin\varphi^4 - 2\sin\varphi^2 \cos\varphi^2 + \cos\varphi^4] d\varphi\, dz = \frac{\lambda_B}{r^3}; \tag{A26}$$

$$K_{\lambda_B[3333]} = \frac{\lambda_B}{\pi r^2 l} \int_0^l \int_0^{2\pi} [0] d\varphi\, dz; \tag{A27}$$

$$K_{\lambda_B[1122]} = \frac{\lambda_B}{\pi r^3 l} \int_0^l \int_0^{2\pi} \left[2\sin \varphi^2 \cos \varphi^2 - \sin \varphi^4 - \cos \varphi^4 \right] d\varphi \, dz = -\frac{\lambda_B}{r^3}; \tag{A28}$$

$$K_{\lambda_B[1133]} = K_{\lambda_B[2233]} = \frac{\lambda_B}{\pi r^3 l} \int_0^l \int_0^{2\pi} [0] d\varphi \, dz = 0; \tag{A29}$$

$$K_{\lambda_B[1212]} = \frac{\lambda_B}{\pi r^3 l} \int_0^l \int_0^{2\pi} \left[4\sin \varphi^2 \cos \varphi^2 \right] d\varphi \, dz = \frac{\lambda_B}{r^3}; \tag{A30}$$

$$K_{\lambda_B[2112]} = \frac{\lambda_B}{\pi r^3 l} \int_0^l \int_0^{2\pi} \left[4\sin \varphi^2 \cos \varphi^2 \right] d\varphi \, dz = \frac{\lambda_B}{r^3}; \tag{A31}$$

$$K_{\lambda_B[1313]} = K_{\lambda_B[2323]} = \frac{\lambda_B}{\pi r^3 l} \int_0^l \int_0^{2\pi} \left[\sin \varphi^2 \right] d\varphi \, dz = \frac{\lambda_B}{r^3}; \tag{A32}$$

$$K_{\lambda_B[3113]} = K_{\lambda_B[3223]} = \frac{\lambda_B}{\pi r^3 l} \int_0^l \int_0^{2\pi} [0] d\varphi \, dz = 0. \tag{A33}$$

As expected, tensor $\mathbf{K}_{\lambda_B}$ is transversely isotropic.

The contribution of surface bending $\hat{\mathbf{C}}_B$ to the stiffness tensor of the equivalent inhomogeneity Equation (35) is

$$\hat{\mathbf{C}}_B = \hat{\mathbf{K}}_{\mu B} + \hat{\mathbf{K}}_{\lambda B}; \tag{A34}$$

$$\hat{C}_{B[11]} = \hat{C}_{B[22]} = \frac{\lambda_B + 2\mu_B}{r^3}, \ \hat{C}_{B[33]} = 0, \tag{A35}$$

$$\hat{C}_{B[12]} = -\frac{\lambda_B + 2\mu_B}{r^3}, \ \hat{C}_{B[13]} = \hat{C}_{B[23]} = 0, \tag{A36}$$

$$\hat{C}_{B[44]} = \hat{C}_{B[55]} = \frac{\mu_B}{r^3}, \ \hat{C}_{B[66]} = \frac{1}{2}\left[\hat{C}_{B[11]} - \hat{C}_{B[12]}\right] = \frac{\lambda_B + 2\mu_B}{r^3}, \tag{A37}$$

In addition to $\hat{\mathbf{C}}_B$, the contribution of surface tension to properties of equivalent inhomogeneity includes another term $\hat{\mathbf{C}}_T$ present in Equation (15); however, evaluation of this term is presented in [42].

References

1. Benveniste, Y.; Miloh, T. Imperfect soft and stiff interfaces in two-dimensional elasticity. *Mech. Mater.* **2001**, *33*, 309–323. [CrossRef]
2. Dong, C.Y.; Zhang, G.L. Boundary element analysis of three dimensional nanoscale inhomogeneities. *Int. J. Solids Struct.* **2013**, *50*, 201–208. [CrossRef]
3. Dong, H.; Wang, J.; Rubin, M. Cosserat interphase models for elasticity with application to the interphase bonding a spherical inclusion to an infinite matrix. *Int. J. Solid. Struct.* **2014**, *51*, 462–477. [CrossRef]
4. Hashin, Z. Thermoelastic properties of fiber composites with imperfect interface. *Mech. Mater.* **1990**, *8*, 333–348. [CrossRef]
5. Hashin, Z. Thermoelastic properties of particulate composites with imperfect interface. *J. Mech. Phys. Solids* **1991**, *39*, 745–762. [CrossRef]
6. He, L.H.; Li, Z.R. Impact of surface stress on stress concentration. *Int. J. Solids Struct.* **2006**, *43*, 6208–6219. [CrossRef]
7. Huang, Z.P.; Wang, J. A theory of hyperelasticity of multi-phase media with surface/interface effect. *Acta Mech.* **2006**, *182*, 195–210. [CrossRef]

8. Jasiuk, I.; Kouider, M.W. The effect of an inhomogeneous interphase on the elastic constants of transversely isotropic composites. *Mech. Mat.* **1993**, *15*, 53–63. [CrossRef]

9. Serpilli, M.; Rizzoni, R.; Lebon, F.; Dumont, S. An asymptotic derivation of a general imperfect interface law for linear multiphysics composites. *Int. J. Solids Struct.* **2019**, *180–181*, 97–107. [CrossRef]

10. Zhang, W.X.; Wang, T. Effect of surface energy on the yield strength of nanoporous materials. *Appl. Phys. Lett.* **2007**, *90*, 063104. [CrossRef]

11. Rubin, M.; Benveniste, Y.A. Cosserat shell model for interphases in elastic media. *J. Mech. Phys. Solids* **2014**, *52*, 1023–1052. [CrossRef]

12. Gurtin, M.E.; Murdoch, A.I. A continuum theory of elastic material surfaces. *Arch. Ration. Mech. Anal.* **1975**, *57*, 291–323. [CrossRef]

13. Gurtin, M.E.; Murdoch, A.I. Surface stress in solids. *Int. J. Solids Struct.* **1978**, *14*, 431–440. [CrossRef]

14. Brisard, S.; Dormieux, L.; Kondo, D. Hashin–Shtrikman bounds on the bulk modulus of a nanocomposite with spherical inhomogeneities and interface effects. *Comput. Mater. Sci.* **2010**, *48*, 589–596. [CrossRef]

15. Chen, T.; Dvorak, G.J. Fibrous nano-composites with interface stresses: Hill's and Levin's connection for effective moduli. *Appl. Phys. Lett.* **2006**, *88*, 211912. [CrossRef]

16. Chen, T.; Dvorak, G.J.; Yu, C.C. Size-dependent elastic properties of unidirectional nano-composites with interface stresses. *Acta Mech.* **2007**, *188*, 39–54. [CrossRef]

17. Duan, H.L.; Wang, J.; Huang, Z.P.; Karihaloo, B.L. Size-dependent effective elastic constants of solids containing nanoinhomogeneities with interface stress. *J. Mech. Phys. Solids* **2005**, *53*, 1574–1596. [CrossRef]

18. Lim, C.W.; Li, Z.R.; He, L.H. Size-dependent, non-uniform elastic field inside a nano-scale spherical inclusion due to interface stress. *Int. J. Solids Struct.* **2006**, *43*, 5055–5065. [CrossRef]

19. Nazarenko, L.; Stolarski, H. Energy-based definition of equivalent inhomogeneity for various interphase models and analysis of effective properties of particulate composites. *Comp. Part B.* **2016**, *94*, 82–94. [CrossRef]

20. Altenbach, H.; Eremeyev, V.A. On the shell theory on the nanoscale with surface stresses. *Int. J. Eng. Sci.* **2011**, *49*, 1294–1301. [CrossRef]

21. Miller, R.E.; Shenoy, V.B. Size-dependent elastic properties of nanosized structural elements. *Nanotechnology* **2000**, *11*, 139–147. [CrossRef]

22. Steigmann, D.J.; Ogden, R.W. Plane deformations of elastic solids with intrinsic boundary elasticity. *Proc. R. Soc. Lond. A* **1997**, *453*, 853–877. [CrossRef]

23. Steigmann, D.J.; Ogden, R.W. Elastic surface-substrate interactions. *Proc. R. Soc. Lond. A* **1999**, *455*, 437–474. [CrossRef]

24. Chhapadia, P.; Mohammadi, P.; Sharma, P. Curvature-dependent surface energy and implications for nanostructures. *J. Mech. Phys. Solids* **2011**, *59*, 2103–2115. [CrossRef]

25. Mohammadi, P.; Sharma, P. Atomistic elucidation of surface roughness on curvature-dependent surface energy, surface stress, and elasticity. *Appl. Phys. Latter* **2012**, *100*, 133110. [CrossRef]

26. Eremeyev, V.A.; Wiczenbach, T. On Effective Bending Stiffness of a Laminate Nanoplate Considering Steigmann–Ogden Surface Elasticity. *Appl. Sci.* **2020**, *10*, 7402. [CrossRef]

27. Dell'Isola, F.; Seppecher, P. Edge contact forces and quasi-balanced power. *Meccanica* **1997**, *32*, 33–52. [CrossRef]

28. Dell'Isola, F.; Seppecher, P.; Madeo, A. How contact interactions may depend on the shape of Cauchy cuts in Nth gradient continua: Approach "à la d'Alembert". *Z. Angew. Math. Phys.* **2012**, *63*, 1119–1141. [CrossRef]

29. Javili, A.; Dell'Isola, F.; Steinmann, P. Geometrically nonlinear higher-gradient elasticity with energetic boundaries. *J. Mech. Phys. Solids* **2013**, *61*, 2381–2401. [CrossRef]

30. Javili, A.; Ottosen, N.S.; Ristinmaa, M.; Mosler, J. Aspects of interface elasticity theory. *Math. Mech. Solids* **2018**, *23*, 1004–1024. [CrossRef]

31. Mindlin, R.D. Micro-structure in linear elasticity. *Arch. Ration. Mech. Anal.* **1964**, *16*, 51–78. [CrossRef]

32. Mindlin, R.D. Second gradient of strain and surface-tension in linear elasticity. *Int. J. Solids Struct.* **1965**, *1*, 417–438. [CrossRef]

33. Toupin, R.A. Elastic materials with couple-stresses. *Arch. Ration. Mech. Anal.* **1962**, *11*, 385–414. [CrossRef]

34. Eremeyev, V.A.; Lebedev, L.P. Mathematical study of boundary-value problems within the framework of Steigmann–Ogden model of surface elasticity. *Contin. Mech. Therm.* **2016**, *28*, 407–422. [CrossRef]

35. Eremeev, V. On dynamic boundary conditions within the linear Steigmann-Ogden model of surface elasticity and strain gradient elasticity. In *Dynamical Processes in Generalized Continua and Structures. Advanced Structured Materials*; Altenbach, H., Belyaev, A., Eremeyev, V., Krivtsov, A., Porubov, A., Eds.; Springer: Berlin/Heidelberg, Germany, 2019; Volume 103.

36. Zemlyanova, A.Y.; Mogilevskaya, S.G. Circular inhomogeneity with Steigmann–Ogden interface: Local fields, neutrality, and Maxwell's type approximation formula. *Int. J. Solids Struct.* **2018**, *135*, 85–98. [CrossRef]

37. Gao, X.; Huang, Z.; Qu, J.; Fang, D. A curvature-dependent interfacial energy-based interface stress theory and its applications to nanostructured materials: (I) general theory. *J. Mech. Phys. Solids* **2014**, *66*, 59–77. [CrossRef]

38. Gao, X.; Huang, Z.; Fang, D. Curvature-dependent interfacial energy and its effects on the elastic properties of nanomaterials. *Int. J. Solid. Struct.* **2017**, *113*, 100–107. [CrossRef]

39. Nazarenko, L.; Stolarski, H.; Altenbach, H. Effective properties of random nano-materials including Steigmann–Ogden interface model of surface. *Comput. Mech..* under review.

40. Zemlyanova, A.Y.; Mogilevskaya, S.G. On spherical inhomogeneity with Steigmann–Ogden interface. *J. Appl. Mech.* **2018**, *85*, 121009. [CrossRef]

41. Nazarenko, L.; Bargmann, S.; Stolarski, H. Energy-equivalent inhomogeneity approach to analysis of effective properties of nano-materials with stochastic structure. *Int. J. Solids Struct.* **2015**, *59*, 183–197. [CrossRef]

42. Nazarenko, L.; Stolarski, H.; Altenbach, H. Effective properties of short-fiber composites with Gurtin-Murdoch model of interphase. *Int. J. Solids Struct.* **2016**, *97–98*, 75–88. [CrossRef]

43. Nazarenko, L.; Bargmann, S.; Stolarski, H. Closed-form formulas for the effective properties of random particulate nanocomposites with complete Gurtin–Murdoch model of material surfaces. *Contin. Mech. Thermodyn.* **2017**, *29*, 77–96. [CrossRef]

44. Itskov, M. *Tensor Algebra and Tensor Analysis for Engineers*; Springer: Berlin/Heidelberg, Germany, 2007.

45. Hill, R. Elastic properties of reinforced solids: Some theoretical principles. *J. Mech. Phys. Solids* **1963**, *11*, 357–372. [CrossRef]

46. Milton, G.W. *The Theory of Composites*; Cambridge University Press: Cambridge, UK, 2002.

47. Christensen, R.M.; Lo, K.M. Solution for effective shear properties in three phase sphere and cylinder models. *J. Mech. Phys. Solids* **1979**, *27*, 315–330. [CrossRef]

Publisher's Note: MDPI stays neutral with regard to jurisdictional claims in published maps and institutional affiliations.

 *technologies*

Article

Asymptotic Justification of Models of Plates Containing Inside Hard Thin Inclusions

Evgeny Rudoy [1,2]

[1] Lavrentyev Institute of Hydrodynamics of SB RAS, 630090 Novosibirsk, Russia; rem@hydro.nsc.ru
[2] Department of Mathematics and Mechanics, Novosibirsk State University, 630090 Novosibirsk, Russia

Received: 1 October 2020; Accepted: 23 October 2020; Published: 28 October 2020

Abstract: An equilibrium problem of the Kirchhoff–Love plate containing a nonhomogeneous inclusion is considered. It is assumed that elastic properties of the inclusion depend on a small parameter characterizing the width of the inclusion ε as ε^N with $N < 1$. The passage to the limit as the parameter ε tends to zero is justified, and an asymptotic model of a plate containing a thin inhomogeneous hard inclusion is constructed. It is shown that there exists two types of thin inclusions: rigid inclusion ($N < -1$) and elastic inclusion ($N = -1$). The inhomogeneity disappears in the case of $N \in (-1, 1)$.

Keywords: Kirchhoff-Love plate; composite material; thin inclusion; asymptotic analysis

1. Introduction

An equilibrium problem of a Kirchhoff–Love plate containing a nonhomogeneous inclusion is considered. It is assumed that the elastic properties of the inclusion depend on a small parameter characterizing width of the inclusion ε as ε^N with $N < 1$. The problem is formulated as a variational one; namely, as a minimization problem of the energy functional over a set of admissible deflections in the Sobolev space H^2. This implies that the deflections function is a solution of a boundary value problem for bi-harmonic operator (pure bending, see, e.g., [1–4]).

The aim of the present work is to justify passing to the limit as $\varepsilon \to 0$. To do this, we apply a method that was originally introduced in [5,6] for problems of gluing plates. The method is based on variational properties of the solution to the corresponding minimization problem and allows for finding a limit problem for any $N < 1$ simultaneously. It is shown that there exist two types of hard inclusions in dependence of N: thin rigid inclusion ($N < -1$) and thin elastic inclusion ($N = -1$). In case $N \in (-1, 1)$, the influence of the inhomogeneity disappears in the limit. We get limit problems in a variational form, which is convenient, for example, for numerical analysis by the finite element method.

Let us give a short survey of works that are close to the present investigation. Note that there are not so many works devoted to study of models of thin inclusions in plates. We mention [7–9], in which thin elastic inclusions in pates were studied. Papers [10–13] are devoted investigations of thin rigid inclusions. We refer to [14–21] for asymptotic analyses for different models of bonded structures in Elasticity. We indicate also paper [22], where a geometry-dependent state problem for a heterogeneous medium with defects is investigated in framework of anti-plane elasticity.

Finally, we mention paper [23], where the mechanical behavior of an anisotropic nonhomogeneous linearly elastic three-layer plate with soft adhesive, including the inertia forces, was studied, and the various limiting models in the dependence of the size and the stiffness of the adhesive was derived. The problem under consideration in the present paper is different from the mentioned paper because we consider the hard inhomogeneity lying strictly inside the plate and derive limiting problem depending on the size and stiffness of the inclusion. Wherein, the plate size does not vary and remains constant.

2. Statement of Problem

Let us fix a small parameter $\varepsilon \in (0,1)$ and consider an inhomogeneous rectangular plate $\Omega \subset \mathbb{R}^2$ with a thin rectangular inclusion $\Omega^\varepsilon_{inc} \subset \Omega$ of width $2\varepsilon d$, where d is diameter of Ω. Let us specify some notations:

$$\Omega = (-a_1, a_2) \times (-b_1, b_2), \ a_\alpha, b_\alpha > 0, \ \alpha = 1, 2,$$

$$\Omega^\varepsilon_{inc} = (-\varepsilon d, \varepsilon d) \times (-c_1, c_2), \ 0 < c_\alpha < b_\alpha, \ \alpha = 1, 2,$$

$$\Omega_\pm = \{(y_1, y_2) \in \Omega \mid \pm y_1 > 0\},$$

$$S = \partial\Omega_- \cap \partial\Omega_+,$$

$$S_{inc} = S \cap \Omega^\varepsilon_{inc},$$

$$\Omega^\varepsilon_{mat} = \Omega \setminus \overline{\Omega}^\varepsilon_{inc}, \ \Omega^\varepsilon_\pm = \Omega^\varepsilon_{mat} \cap \Omega_\pm,$$

Note that, for all small enough $\varepsilon > 0$ a family of subdomains Ω^ε_{inc} lies strictly inside Ω. Besides, let us define the following notations:

$$\Omega^\varepsilon_{mid} = \{(y_1, y_2) \in \Omega \mid -\varepsilon d < y_1 < \varepsilon d, \ y_2 \in S\},$$

$$S^\varepsilon_\pm = \{(y_1, y_2) \in \Omega \mid y_1 = \pm\varepsilon d, \ y_2 \in S\},$$

We assume that S_{inc} is divided into three subsets $S_\alpha \subset S_{inc}$, where each S_α is an union of finite number of segments or empty set, $\alpha = 1, 2, 3$.

In our consideration, Ω is a composite plate, consisting of the elastic matrix Ω^ε_{mat} and the inhomogeneous inclusion $\Omega^\varepsilon_{inc} = \cup^3_{\alpha=1} \Omega^\varepsilon_\alpha$, where

$$\Omega^\varepsilon_\alpha = \{(y_1, y_2) \in \mathbb{R}^2 \mid -\varepsilon d < y_1 < \varepsilon d, \ y_2 \in S_\alpha\}, \ \alpha = 1, 2, 3.$$

Moreover, in the sequel, we will use the following notations:

$$\Omega^\varepsilon_0 = \Omega^\varepsilon_{mid} \setminus \cup^3_{\alpha=1} \overline{\Omega}^\varepsilon_\alpha,$$

$$S_0 = S \setminus \overline{S}_{inc}.$$

Denote, by E_0, E^ε_α and k_0, k_α, Young's modules and Poisson's ratios of parts Ω_{mat} and $\Omega^\varepsilon_\alpha$ of the composite plate Ω, respectively, $\alpha = 1, 2, 3$. The compound character of the structure is expressed by the fact that E_0, k_0, and k_α are constants, while Young's modulus E^ε_α depends on ε, as follows:

$$E^\varepsilon_\alpha = \varepsilon^{N_\alpha} E_\alpha \ \text{in} \ \Omega^\varepsilon_\alpha, \ \alpha = 1, 2, 3,$$

where N_1, N_2, N_3 are real numbers, such that

$$N_1 < -1, \ N_2 = -1, \ N_3 \in (-1, 1).$$

Parameters N_1 and N_2 correspond to hard inclusions in the plate Ω (see [6,24,25]). Moreover, put $N_0 = 0$.

Denote, by w, deflections of the composite plate Ω. Then the bending moments are defined by formulae (see, e.g., [26,27])

$$m_{ij}(w) = d^\varepsilon_{ijkl} w_{,kl}, \ i, j = 1, 2, \ w_{,kl} = \frac{\partial^2 w}{\partial y_k \partial y_l},$$

where the positive definite and symmetric tensor $\{d_{ijkl}\}$ is orthotropic with the following components:

$$d^\varepsilon_{iiii}(y) = D^\varepsilon(y), \ d^\varepsilon_{iijj}(y) = D^\varepsilon(y)k^\varepsilon(y),$$
$$d^\varepsilon_{ijij}(y) = d^\varepsilon_{ijji}(y) = D^\varepsilon(y)(1 - k^\varepsilon(y))/2, \ i \neq j, \ i,j = 1,2, \tag{1}$$

$$D^\varepsilon(y) = \begin{cases} D_0 & \text{in } \Omega^\varepsilon_{mat}, \\ \varepsilon^{N_\alpha} D_\alpha & \text{in } \Omega^\varepsilon_\alpha, \ \alpha = 1,2,3, \end{cases}$$

$$D_\alpha = \frac{E_\alpha h^3}{12(1 - k^2_\alpha)}, \ \alpha = 0,1,2,3,$$

$$k^\varepsilon(y) = \begin{cases} k_0 & \text{in } \Omega^\varepsilon_{mat}, \\ k^\varepsilon_\alpha, & \text{in } \Omega^\varepsilon_\alpha, \ \alpha = 1,2,3, \end{cases}$$

h is a thickness of the plate Ω that is constant. Note paper [28], where it was shown non-standard behaviour in the asymptotic two-dimensional reduction from three-dimensional elasticity, when the thickness and size of inclusions depend on the same parameter.

The potential energy functional of the plate has the following representation (see [27]):

$$\Pi(w) = \frac{1}{2} \int_\Omega d^\varepsilon_{ijkl} w_{,kl} w_{,ij} \, dy - \int_\Omega fw \, dy,$$

where $f \in L_2(\Omega)$ is a bulk force acting on the plate Ω. Subsequently, the equilibrium problem of nonhomogeneous plate clamped on the external boundary $\partial\Omega$ can be formulated as the minimization problem: find a function $w_\varepsilon \in H^2_0(\Omega)$ such that

$$\Pi(w_\varepsilon) = \inf_{w \in H^2_0(\Omega)} \Pi(w). \tag{2}$$

Problem (2) is known to have a unique solution w_ε (see, e.g., [26,29]), which satisfies the variational equality:

$$\int_\Omega d^\varepsilon_{ijkl} w_{\varepsilon,kl} w_{,ij} \, dy = \int_\Omega fw \, dy \quad \forall w \in H^2_0(\Omega). \tag{3}$$

Moreover, the function w_ε is a unique solution the following boundary value problem:

$$(d^\varepsilon_{ijkl} w_{\varepsilon,kl})_{,ij} = f \ \text{ in } \Omega,$$

$$w_\varepsilon = \frac{\partial w_\varepsilon}{\partial \nu} = 0 \ \text{ on } \partial\Omega,$$

where ν is a unit normal vector $\partial\Omega$.

3. Decomposition of the Problem and Coordinate Transformations

In the sequel, we will have deal with the problem (3). Let us rewrite it in an equivalent form. For this, we introduce the following set:

$$K_\varepsilon = \{v = (v_-, v_+, v_m) \in H^2(\Omega^\varepsilon_-) \times H^2(\Omega^\varepsilon_+) \times H^2(\Omega^\varepsilon_m) \ |$$

$$v_\pm = v_m, \ v_{\pm,1} = v_{m,1} \ \text{a.e. on } S^\varepsilon_\pm,$$

$$v_\pm = \frac{\partial v_\pm}{\partial \nu} = 0 \ \text{a.e. on } \partial\Omega^\varepsilon_\pm \cap \partial\Omega\}.$$

Taking into account the (1), problem (3) can be reformulated, as follows: find a triplet $(w_{\varepsilon-}, w_{\varepsilon+}, w_{\varepsilon m}) \in K_\varepsilon$ satisfying a variational equality

$$b_{\varepsilon-}(w_{\varepsilon-}, v_-) + b_{\varepsilon+}(w_{\varepsilon+}, v_+) + b_{\varepsilon m}(w_{\varepsilon m}, v_m) =$$
$$= l_-(v_-) + l_+(v_+) + l_m(v_m) \quad \forall (v_-, v_+, v_m) \in K_\varepsilon, \tag{4}$$

where

$$b_{\varepsilon\pm}(u, v) = D_0 \int_{\Omega_\pm^\varepsilon} (u_{,11}v_{,11} + u_{,22}v_{,22} + k_0(u_{,11}v_{,22} + u_{,22}v_{,11}) + 2(1 - k_0)u_{,12}v_{,12}) \, dy,$$

$$b_{\varepsilon m}(u, v) = \sum_{\alpha=0}^{3} D_\alpha^\varepsilon \int_{\Omega_\alpha^\varepsilon} (u_{,11}v_{,11} + u_{,22}v_{,22} + k_\alpha(u_{,11}v_{,22} + u_{,22}v_{,11}) + 2(1 - k_\alpha)u_{,12}v_{,12}) \, dy.$$

$$l_{\varepsilon\pm}(u) = \int_{\Omega_\pm^\varepsilon} fu \, dy, \quad l_{\varepsilon m}(u) = \int_{\Omega_m^\varepsilon} fu \, dy.$$

From the Calculus of Variations, it follows that problem (4) has a unique solution $(w_{\varepsilon-}, w_{\varepsilon+}, _{\varepsilon m}) \in K_\varepsilon$ for all $\varepsilon > 0$ small enough (see, e.g., [2,26]). Herewith, $w_{\varepsilon\pm}$ and $w_{\varepsilon m}$ are restrictions of w_ε on subdomains $\Omega_\pm^\varepsilon$ and Ω_m^ε, respectively.

Next, we introduce coordinate transformations that map domains $\Omega_\pm^\varepsilon$ and Ω_m^ε onto domains independent of ε. For this, we consider two convex domains ω_1 and ω_2, such that

$$\overline{S} \subset \omega_1, \ \overline{\omega}_1 \subset \omega_2, \ \partial\omega_2 \cap \{y_1 = -a_1\} = \emptyset, \ \partial\omega_2 \cap \{y_1 = a_2\} = \emptyset,$$

and a smooth cut-off function θ, such that

$$\theta = 1 \text{ in } \overline{\omega}_1, \ 0 < \theta < 1 \text{ in } \omega_2, \ \theta = 0 \text{ in } \mathbb{R}^2 \setminus \overline{\omega}_2.$$

Let us introduce the following notations:

$$\Omega_m = \{(z_1, z_2) \in \mathbb{R}^2 \mid -d < z_1 < d, \ z_2 \in S\},$$

$$S_\pm = \{(z_1, z_2) \in \mathbb{R}^2 \mid z_1 = \pm d, \ z_2 \in S\},$$

$$\Omega_\alpha = \{(z_1, z_2) \in \mathbb{R}^2 \mid -d < z_1 < d, \ z_2 \in S_\alpha\}, \ \alpha = 0, 1, 2, 3,$$

$$S_\alpha^\pm = \{(z_1, z_2) \in \mathbb{R}_z^2 \mid z_1 = \pm d, \ z_2 \in S_\alpha\}, \ \alpha = 0, 1, 2, 3.$$

and define coordinate transformations in the domains $\Omega_\pm$ and Ω_m as follows:

$$y_1 = x_1 \pm \varepsilon d\theta(x_1, x_2), \quad y_2 = x_2, \quad (x_1, x_2) \in \Omega_\pm, \quad (y_1, y_2) \in \Omega_\pm^\varepsilon, \tag{5}$$

$$y_1 = \varepsilon z_1, \quad y_2 = z_2, \quad (z_1, z_2) \in \Omega_m, \quad (y_1, y_2) \in \Omega_m^\varepsilon. \tag{6}$$

It is not difficult to show that for all sufficiently small coordinate transformations (5) and (6) map bijectively the domains $\Omega_\pm$ and Ω_m onto $\Omega_\pm^\varepsilon$ and Ω_m^ε, respectively, (see, e.g., [30,31]). Note that the subdomain $\Omega_\alpha^\varepsilon$ is mapped into subdomains Ω_α, $\alpha = 0, 1, 2, 3$.

Denote, by $\Phi_\varepsilon^\pm(x)$ and $J_\varepsilon^\pm$, Jacobian matrices and Jacobians of transformations (5), respectively,

$$\Phi_\varepsilon^\pm(x_1, x_2) = \begin{pmatrix} 1 \pm \varepsilon d\theta_{,1}(x_1, x_2) & \pm\varepsilon d\theta_{,2}(x_1, x_2) \\ 0 & 1 \end{pmatrix},$$

$$J_\varepsilon^\pm(x_1, x_2) = \det \Phi_\varepsilon^\pm(x_1, x_2) = 1 \pm \varepsilon d\theta_{,1}(x_1, x_2).$$

Coordinate transformations (5) and (6) establish one-to-one correspondences between spaces $H^2(\Omega_\pm)$, $H^2(\Omega_m)$ and $H^2(\Omega_\pm^\varepsilon)$, $H^2(\Omega_m^\varepsilon)$, respectively. Moreover, the set K^ε is transformed into a set K_ε,

$$K_\varepsilon = \{v = (v_-, v_+, v_m) \in H^{2,0}(\Omega_-) \times H^{2,0}(\Omega_+) \times H^2(\Omega_m) \ \mid$$

$$v_\pm|_S = v_m|_{S_\pm}, \quad v_{\pm,1}|_S = \frac{1}{\varepsilon} v_{m,1}|_{S_\pm} \},$$

where

$$H^{2,0}(\Omega_\pm) = \{v_\pm \in H^2(\Omega_\pm) \ \mid \ v_\pm = \frac{\partial v_\pm}{\partial \nu} = 0 \text{ a.e. on } \partial\Omega_\pm^\varepsilon \cap \partial\Omega\}.$$

Hereinafter, we assume that, for any functions $v_\pm(x)$, $x \in \Omega_\pm$, and $v_m(z)$, $z \in \Omega_m$, equality $v_\pm|_S = v_m|_{S_\pm}$ means that

$$v_\pm(0, x_2) = v_m(\pm d, z_2), \quad x_2 = z_2 \in S.$$

Introduce the following notations:

$$w_\pm^\varepsilon(x_1, x_2) = w_{\varepsilon\pm}(x_1 \pm \varepsilon d\theta(x_1, x_2), x_2), \quad (x_1, x_2) \in \Omega_\pm,$$

$$w_m^\varepsilon(z_1, z_2) = w_{\varepsilon m}(\varepsilon z_1, z_2), \quad (z_1, z_2) \in \Omega_m.$$

Becase of the smoothness of coordinate transformations (5), we have asymptotic expansions for the transformations of the second-order derivatives for (5) (see, e.g., [30–33])

$$w_{\varepsilon\pm,ij} = w_{\pm,ij}^\varepsilon + \varepsilon P_{ij}^\pm(\varepsilon, w_\pm^\varepsilon), \tag{7}$$

with

$$|P_{ij}^\pm(\varepsilon, w_\pm^\varepsilon)| \leq C(|w_{\pm,k}^\varepsilon| + |w_{\pm,kl}^\varepsilon|), \quad i,j,k,l = 1,2.$$

Besides, we have for (6)

$$w_{\varepsilon m,11}(y_1, y_2) = \frac{w_{m,11}^\varepsilon(z_1, z_2)}{\varepsilon^2}, \quad w_{\varepsilon m,12}(y_1, y_2) = \frac{w_{m,12}^\varepsilon(z_1, z_2)}{\varepsilon}, \quad w_{\varepsilon m,22}(y_1, y_2) = w_{m,22}^\varepsilon(z_1, z_2).$$

After applying coordinate transformations (5) and (6) to (4), we get that the triplet $(w_-^\varepsilon, w_+^\varepsilon, uw_m^\varepsilon) \in K_\varepsilon$ is a unique solution to the following variational equality:

$$b_-^\varepsilon(w_-^\varepsilon, v_-) + b_+^\varepsilon(w_+^\varepsilon, v_+) + b_m^\varepsilon(w_m^\varepsilon, v_m) = l_-^\varepsilon(v_-) + l_+^\varepsilon(v_+) + l_m^\varepsilon(v_m) \ \forall (v_-, v_+, v_m) \in K_\varepsilon, \tag{8}$$

where, taking into account (7) and (1),

$$b_\pm^\varepsilon(u, v) - b_\pm(u, v) + r_\pm(\varepsilon, u, v),$$

$$b_\pm(u, v) = D_0 \int_{\Omega_\pm} (u_{,11}v_{,11} + u_{,22}v_{,22} + k_\pm(u_{,11}v_{,22} + u_{,22}v_{,11}) + 2(1 - k_\pm)u_{,12}v_{,12}) \, dx,$$

$$|r_\pm(\varepsilon, u, v)| \leq c_\pm(\varepsilon) \left(\|u\|^2_{H^2(\Omega_\pm)} + \|v\|^2_{H^2(\Omega_\pm)} \right), \ 0 \leq c_\pm(\varepsilon) = o(1) \text{ as } \varepsilon \to 0, \tag{9}$$

$$b_m^\varepsilon(u,v) =$$

$$= D_0 \int_{\Omega_m} \left(\frac{u_{,11}v_{,11}}{\varepsilon^3} + \varepsilon u_{,22}v_{,22} + \frac{k_m}{\varepsilon}(u_{,11}v_{,22} + v_{,22}w_{,11}) + \frac{2(1-k_m)}{\varepsilon}u_{,12}v_{,12} \right) dz +$$

$$+ D_1 \int_{\Omega_m} \left(\frac{u_{,11}v_{,11}}{\varepsilon^{3-N_1}} + \frac{u_{,22}v_{,22}}{\varepsilon^{-N_1-1}} + \frac{k_m}{\varepsilon^{1-N_1}}(u_{,11}v_{,22} + v_{,22}w_{,11}) + \frac{2(1-k_m)}{\varepsilon^{1-N_1}}u_{,12}v_{,12} \right) dz +$$

$$+ D_2 \int_{\Omega_m} \left(\frac{u_{,11}v_{,11}}{\varepsilon^4} + u_{,22}v_{,22} + \frac{k_m}{\varepsilon^2}(u_{,11}v_{,22} + v_{,22}w_{,11}) + \frac{2(1-k_m)}{\varepsilon^2}u_{,12}v_{,12} \right) dz +$$

$$+ D_3 \int_{\Omega_m} \left(\frac{u_{,11}v_{,11}}{\varepsilon^{3-N_3}} + \varepsilon^{N_3+1}u_{,22}v_{,22} + \frac{k_m}{\varepsilon^{1-N_3}}(u_{,11}v_{,22} + v_{,22}w_{,11}) + \frac{2(1-k_m)}{\varepsilon^{1-N_3}}u_{,12}v_{,12} \right) dz,$$

$$l_\pm^\varepsilon(v) = \int_{\Omega_\pm} f(x_1 \pm d\theta(x_1,x_2), x_2)(1 \pm d\theta_{,1}(x_1,x_2)v\,dx,$$

$$l_m^\varepsilon(v) = \varepsilon \int_{\Omega_m} f(\varepsilon z_1, z_2)v\,dz,$$

$$|l_\pm^\varepsilon(v)| \leq C\|v\|_{L_2(\Omega_\pm)}, \tag{10}$$

$$|l_m^\varepsilon(v)| \leq C\varepsilon\|v\|_{L_2(\Omega_m)}. \tag{11}$$

4. Limit Problem

To justify passing to the limit as $\varepsilon \to 0$, we need some auxiliary lemma proved in [5,6].

Lemma 1 (Poincare-typé inequalities). *For any triplet $(v_-, v_+, v_m) \in K_\varepsilon$ and $\varepsilon \in (0,1)$, the inequalities*

$$\|v_m\|_{L_2(\Omega_m)}^2 \leq C\left(\|v_{m,11}\|_{L_2(\Omega_m)}^2 + \|v_\pm\|_{H^{2,0}(\Omega_\pm)}^2\right),$$

$$\|v_{m,1}\|_{L_2(\Omega_m)}^2 \leq C\left(\|v_{m,11}\|_{L_2(\Omega_m)}^2 + \varepsilon^2\|v_{\pm,1}\|_{L_2(S)}^2\right)$$

hold, where a constant $C > 0$ does not depend on (v_-, v_+, v_m) and $\varepsilon > 0$.

Our main result is the following theorem.

Theorem 1. *Let $w^\varepsilon = (w_-^\varepsilon, w_+^\varepsilon, w_m^\varepsilon)$ be a solution to (8); let $w_0 \in K_0$ be a solution to the following variational equality:*

$$b(w_0, w) + 4d(1-k_2)D_m \int_{S_2} \frac{\partial(w_{0,1}|_{S_2})}{\partial z_2}\frac{\partial(w_{,1}|_{S_2})}{\partial z_2}\,dz_2 = l(w)\ \forall w \in K_0, \tag{12}$$

where

$$K_0 = \{w \in H_0^2(\Omega) \mid w = \alpha x_2 + \beta \text{ a.e. on } S_1,\ \alpha, \beta \in \mathbb{R};\ w_{,1} \in H^1(S_2)\}.$$

Denote, by $w_\pm$, a restriction of w to subdomain $\Omega_\pm$ and, moreover, put

$$w_m(z_1, z_2) = w_0(z_1, 0) \quad \text{for} \quad (z_1, z_2) \in \Omega_m.$$

Then, the following convergences

$$w_\pm^\varepsilon \rightharpoonup w_\pm \quad \text{weakly in} \quad H^2(\Omega_\pm),$$

$$w_m^\varepsilon \rightharpoonup w_m \quad \text{weakly in} \quad L_2(\Omega_m),$$

take place as $\varepsilon \to 0$.

Proof. Let us substitute $(w_-^\varepsilon, w_+^\varepsilon, w_m^\varepsilon)$ in (8) as a test function. Taking into account Lemma, (9)–(11), we obtain an estimate

$$\|w_-^\varepsilon\|_{H^{2,0}(\Omega_-)}^2 + \|w_+^\varepsilon\|_{H^{2,0}(\Omega_+)}^2 +$$

$$+ \left\|\frac{w_{m_0,11}^\varepsilon}{\varepsilon^{\frac{3}{2}}}\right\|_{L_2(\Omega_0)}^2 + \left\|\frac{w_{m_0,12}^\varepsilon}{\varepsilon^{\frac{1}{2}}}\right\|_{L_2(\Omega_0)}^2 + \|\varepsilon^{\frac{1}{2}} w_{m_0,22}^\varepsilon\|_{L_2(\Omega_0)}^2 +$$

$$+ \left\|\frac{w_{m_1,11}^\varepsilon}{\varepsilon^{\frac{3-N_1}{2}}}\right\|_{L_2(\Omega_1)}^2 + \left\|\frac{w_{m_1,12}^\varepsilon}{\varepsilon^{\frac{1-N_1}{2}}}\right\|_{L_2(\Omega_1)}^2 + \left\|\frac{w_{m_1,22}^\varepsilon}{\varepsilon^{\frac{-N_1-1}{2}}}\right\|_{L_2(\Omega_1)}^2 +$$

$$+ \left\|\frac{w_{m_2,11}^\varepsilon}{\varepsilon^2}\right\|_{L_2(\Omega_2)}^2 + \left\|\frac{w_{m_2,12}^\varepsilon}{\varepsilon}\right\|_{L_2(\Omega_2)}^2 + \|w_{m_2,22}^\varepsilon\|_{L_2(\Omega_2)}^2 +$$

$$+ \left\|\frac{w_{m_3,11}^\varepsilon}{\varepsilon^{\frac{3-N_3}{2}}}\right\|_{L_2(\Omega_3)}^2 + \left\|\frac{w_{m_3,12}^\varepsilon}{\varepsilon^{\frac{1-N_3}{2}}}\right\|_{L_2(\Omega_3)}^2 + \|\varepsilon^{\frac{N_3+1}{2}} w_{m_3,22}^\varepsilon\|_{L_2(\Omega_3)}^2 \le C \quad (13)$$

with a constant C independent of ε. Here, by $w_{m_\alpha}^\varepsilon$, denote a restriction of w^ε to Ω_α, $\alpha = 0, 1, 2, 3$. Moreover, from (13), Lemma, and definition of the set K_ε, we additionally have

$$\|w_m^\varepsilon\|_{L_2(\Omega_m)} \le C, \quad \|w_{m,1}^\varepsilon\|_{L_2(\Omega_m)} \le C\varepsilon. \quad (14)$$

Estimates (13) and (14) entail the existence of functions $w_\pm \in H^{2,0}(\Omega_\pm)$, $w_m \in L_2(\Omega_m)$, $p_\alpha, q_\alpha, r_\alpha \in L_2(\Omega_\alpha)$, $\alpha = 0, 1, 2, 3$, such that for some subsequence $\{\varepsilon_n\}_{n=1}^\infty$ still denoted by ε, the following convergences:

$$\begin{aligned}
w_\pm^\varepsilon &\rightharpoonup w_\pm \quad \text{weakly in} \quad H^2(\Omega_\pm), \\
w_m^\varepsilon &\rightharpoonup w_m \quad \text{weakly in} \quad L_2(\Omega_\alpha), \\
\varepsilon^{\frac{N_\alpha-3}{2}} w_{m,11}^\varepsilon &\rightharpoonup p_\alpha \quad \text{weakly in} \quad L_2(\Omega_\alpha), \\
\varepsilon^{\frac{N_\alpha-1}{2}} w_{m,12}^\varepsilon &\rightharpoonup q_\alpha \quad \text{weakly in} \quad L_2(\Omega_\alpha), \\
\varepsilon^{\frac{N_\alpha+1}{2}} w_{m,22}^\varepsilon &\rightharpoonup r_\alpha \quad \text{weakly in} \quad L_2(\Omega_\alpha)
\end{aligned} \quad (15)$$

hold as $\varepsilon \to 0$, with $r_2 = w_{m,22}$. Moreover, from (13) and (14), it follows that

$$w_{m,1}^\varepsilon \to w_{m,1} = 0 \quad \text{strongly in} \quad L_2(\Omega_m), \quad (16)$$

$$w_{m,11}^\varepsilon \to w_{m,11} = 0 \quad \text{strongly in} \quad L_2(\Omega_m), \quad (17)$$

$$w_{m,22}^\varepsilon \to w_{m,22} = 0 \quad \text{strongly in} \quad L_2(\Omega_1), \quad (18)$$

and there exists $u \in L_2(\Omega_{m_2})$ such that

$$\frac{w^\varepsilon}{\varepsilon} \rightharpoonup u \quad \text{weakly in} \quad L_2(\Omega_2).$$

From definition of the set K_ε, after passing to the limit as $\varepsilon \to 0$, we obtain

$$w_m|_{S_\pm} = w_\pm|_S. \quad (19)$$

Because $w_{m,1} = 0$ in Ω_m (see (16)), w_m does not depend on z_2. Therefore, taking into account (17), we conclude that there exists a function $\beta(z_2) \in L_2(\Omega_m)$ such that

$$w_m(z_1, z_2) = \beta(z_2), \quad (z_1, z_2) \in \Omega_m.$$

Condition (18) means that the function w_m is affine in the domain Ω_m with respect to z_2, i.e., there exists $\delta, \gamma \in \mathbb{R}$, such that

$$w_m(z_1, z_2) = \delta z_2 + \gamma \quad \text{in} \quad \Omega_1. \tag{20}$$

Because of (19), we have

$$w_-|_S = w_+|_S. \tag{21}$$

Now, let us show that $w_\pm$ satisfy the following equality:

$$w_{+,1} = w_{-,1} \quad \text{on} \quad S. \tag{22}$$

Indeed, from the relation

$$\int_{-d}^{d} w^\varepsilon_{m,11}(z_1, z_2)\, dz_1 = w^\varepsilon_{m,1}(d, z_2) - w^\varepsilon_{m,1}(-d, z_2),$$

it follows that

$$\int_a^b |w^\varepsilon_{m,1}(d, z_2) - w^\varepsilon_{m,1}(-d, z_2)|^2\, dz_2 \le 2d \|w^\varepsilon_{m,11}\|^2_{L_2(\Omega_m)}.$$

Due to estimate (13) and the equalities $w^\varepsilon_{m,1}(\pm d, z_2) = \varepsilon w^\varepsilon_\pm(0, z_2)$ for $z_2 \in (a, b)$ (see the definition of the set K_ε), we obtain

$$\|w^\varepsilon_{+,1} - w^\varepsilon_{-,1}\|_{L_2(S)} \le \frac{2d}{\varepsilon} \|w^\varepsilon_{m,11}\|_{L_2(\Omega_m)} \to 0$$

as $\varepsilon \to 0$. From (15) (the first line) and the compactness of trace operator, it follows

$$w^\varepsilon_{\pm,1} \to w_{\pm,1} \quad \text{strongly in} \quad L_2(S)$$

as $\varepsilon \to 0$, and (22) holds.

At last, using the same arguments as in [6], we can prove additionally that

$$w_{\pm,1}|_{S_2} \in H^1(S_2) \tag{23}$$

and, moreover,

$$p_2 = -k_m w_{m,22} \quad \text{in} \quad \Omega_2,$$

$$q_2 = \frac{\partial(w_{-,1}|_{S_2})}{\partial z_2} \quad \text{in} \quad \Omega_2,$$

$$u = w_{-,1}|_{S_2} \quad \text{in} \quad \Omega_2.$$

Now, let us define a function

$$w_0(x) = \begin{cases} w_-(x) & x \in \Omega_-, \\ w_+(x) & x \in \Omega_+. \end{cases} \tag{24}$$

Conditions (19)–(23) imply that the function w_0 belongs to the set K_0.

In order to proceed with a problem defining the function w_0, we take arbitrary function $v \in C^2(\Omega) \cap K_0$ and define three functions v_-, v_+, v_m by

$$v_- = v|_{\Omega_-}, \quad v_+ = v|_{\Omega_+},$$

$$v_m(z_1, z_2) = v(0, z_2), \quad (z_1, z_2) \in \Omega_m.$$

Subsequently, for these functions, we consider a triplet $(v_- + \varepsilon\psi_-, v_+ + \varepsilon\psi_+, v_m + \varepsilon\psi_m) \in K_\varepsilon$, where $\psi_m(z_1, z_2) = v_{,1}(0, z_2)z_1$ for $(z_1, z_2) \in \Omega_m$, and $\psi_\pm \in H^{2,0}(\Omega_\pm)$ is arbitrary extensions of ψ_m in domains $\Omega_\pm$, such that

$$\psi_\pm|_S = \psi_m|_{S_m^\pm}, \quad \psi_{\pm,1} = 0 \text{ on } S,$$

and substitute it in (8). Since $v_{m,11} = 0$ and $\psi_{m,11} = 0$ in Ω_m, weak convergences in (15) and Formulas (23) allows for us to pass to the limit as $\varepsilon \to 0$ and obtain the following relation:

$$b_-(w_-, v_-) + b_+(w_+, v_+) + 4d(1 - k_2)D_2 \int\limits_{S_2} \frac{\partial(w_{-,1}|_{S_2})}{\partial z_2} \frac{\partial(v_{-,1}|_{S_2})}{\partial z_2} dz_2 =$$

$$= l_-(v_-) + l_+(v_+) \quad \forall v \in C^2(\Omega) \cap K_0.$$

Taking into account (24) and the fact that $C^2(\Omega) \cap K_0$ is dense in K_0, we obtain (12). $\square$

Assuming that the solution w_0 to variational problem (12) has additional regularity, by applying the generalized Green formula (see, e.g., [2,26]), we deduce differential equations and boundary conditions for the functions w_0:

$$D_0\Delta^2 w_0 = f \text{ in } \Omega \setminus (\overline{S}_1 \cup \overline{S}_2),$$

$$w_0 = \frac{\partial w_0}{\partial v} = 0 \text{ on } \partial\Omega,$$

$$w_0 = \delta_0 x_2 + \beta_0 \text{ on } S_1, \quad \delta_0, \beta_0 \in \mathbb{R},$$

$$[m^1(w_0)] = 0 \text{ on } S_1,$$

$$\int\limits_{S_1} [t^1(w_0)] dx_2 = 0, \quad \int\limits_{S_1} [t^1(w_0)]x_2 dx_2 = 0,$$

$$[t^2(w_0)] = 0 \text{ on } S_2,$$

$$p = w_{0,1} \text{ on } S_2,$$

$$4dD_2(1 - k_2)p_{,22} = [m^2(w_0)] \text{ on } S_2,$$

$$p_{,2} = 0 \text{ at } \partial S_2,$$

where $m^\alpha(w_0)$ and $t^\alpha(w_0)$ are bending moments and transverse forces, respectively, defined by

$$m^\alpha(w_0) = D_\alpha\left(k_\alpha\Delta w_0 + (1 - k_\alpha)\frac{\partial^2 w_0}{\partial v^2}\right),$$

$$t^\alpha(w_0) = D_\alpha\frac{\partial}{\partial v}\left(\Delta w_0 + (1 - k_\alpha)\frac{\partial^2 w_0}{\partial \tau^2}\right),$$

$v = (1,0)$ and $\tau = (-1,0)$ are an unit normal vector and an unit tangent vector, respectively, $\alpha = 1, 2$. The mechanical interpretation of boundary conditions can be found in [6], see also [10,34,35].

5. Concluding Remarks

We proposed a method of asymptotic derivation of plate models containing hard thin inclusions lying strictly inside the plate. The method is based on the variational properties of the solution of the equilibrium problem and allows for one to simultaneously construct all possible cases of hard thin inclusions. It is shown that there exist two type of thin inclusions in the Kirchhoff–Love plate, namely, the rigid inclusion S_1 for $N < -1$ and the elastic inclusion S_2 for $N = -1$. The inhomogeneity disappears in the case of $N \in (-1, 1)$. The last means that we have no any peculiarity along the set S_3.

In the conclusion, we note that the proposed method does not allow considering the case of the exponent $N \geq 1$ simultaneously with the case of the exponent $N < 1$, because, for the first case, we need to use other type of test functions (see [6]), which cannot be substituted in variational equality for the second case of the exponent.

Funding: The work is supported by the Russian Foundation for Basic Research (project 18-29-10007).

Conflicts of Interest: The authors have not conflict of interest.

References

1. Destuynder P.; Salaun M. *Mathematical Analysis of Thin Plate Models*; Springer: Berlin/ Heidelberg, Germany, 1996.
2. Khludnev A.M.; Sokolowski J. *Modelling and Control in Solid Mechanics*; Birkhäuser: Basel, Switzerland, 1997.
3. Lagnese J.E.; Lions J.-L. *Modelling Analysis and Control of Thin Plates*; MASSON: Paris, France; Milan, France; Barcelone, Spain, 1988
4. Sweers G. A survey on boundary conditions for the biharmonic. *Complex Var. Elliptic Equ.* **2009**, *54*, 79–93. [CrossRef]
5. Rudoy E.M. Asymptotic modelling of bonded plates by a soft thin adhesive layer. *Sib. Electron. Math. Rep.* **2020**, *17*, 615–625. [CrossRef]
6. Furtsev A.; Rudoy E. Variational approach to modeling soft and stiff interfaces in the Kirchhoff-Love theory of plates. *Int. J. Solids Struct.* **2020**, *202*, 562–574. [CrossRef]
7. Kunets, Y.I.; Matus V.V. Modeling of flexural vibrations of a Kirchhoff plate with a thin-walled elastic inclusion of weak contrast. *J. Math. Sci.* **2012**, *187*, 667–674. [CrossRef]
8. Furtsev A.I. On Contact Between a Thin Obstacle and a Plate Containing a Thin Inclusion. *J. Math. Sci.* **2019**, *237*, 530–545. [CrossRef]
9. Furtsev A.I. The unilateral contact problem for a Timoshenko plate and a thin elastic obstacle. *Sib. Electron. Math. Rep.* **2020**, *17*, 364–379. [CrossRef]
10. Khludnev A. Thin rigid inclusions with delaminations in elastic plates. *Eur. J. Mech. A/Solids* **2011**, *32*, 69–75. [CrossRef]
11. Khludnev A.M. On bending an elastic plate with a delaminated thin rigid inclusion. *J. Appl. Ind. Math.* **2011**, *5*, 582–594. [CrossRef]
12. Fankina I.V. A contact problem for an elastic plate with a thin rigid inclusion. *J. Appl. Ind. Math.* **2016**, *10*, 333–340. [CrossRef]
13. Shcherbakov V.V. Existence of an optimal shape of the thin rigid inclusions in the Kirchhoff-Love plate. *J. Appl. Ind. Math.* **2014**, *8*, 97–105. [CrossRef]
14. Dumont S.; Lebon F.; Rizzoni R. Imperfect interfaces with graded materials and unilateral conditions: theoretical and numerical study. *Math. Mech. Solids* **2018**, *23*, 445–460. [CrossRef]
15. Serpilli M.; Rizzoni R.; Lebon F.; Dumont S. An asymptotic derivation of a general imperfect interface law for linear multiphysics composites. *Int. J. Solids Struct.* **2019**, *180–181*, 97–107. [CrossRef]
16. Dumont S.; Rizzoni R.; Lebon F.; Sacco E. Soft and hard interface models for bonded elements. *Compos. Part Eng.* **2018**, *153*, 480–490. [CrossRef]
17. Schmidt P. Modelling of adhesively bonded joints by an asymptotic method. *Int. J. Eng. Sci.* **2008**, *46*, 1291–1324. [CrossRef]
18. Serpilli M.; Rizzoni R.; Dumont S.; Lebon F. Interface laws for multi-physic composites. In *Proceedings of XXIV AIMETA Conference 2019*; Lecture Notes in Mechanical Engineering; Carcaterra A., Paolone A., Graziani G., Eds.; Springer: Cham, Switzerland, 2020.

19. Bonaldi F.; Geymonat G.; Krasucki F.; Serpilli M. An asymptotic plate model for magneto-electro-thermo-elastic sensors and actuators. *Math. Mech. Solids* **2017**, *22*, 1–25. [CrossRef]
20. Serpilli M. On modeling interfaces in linear micropolar composites. *Math. Mech. Solids* **2018**, *23*, 667–685. [CrossRef]
21. Raffa M.L.; Lebon F.; Rizzoni R. Derivation of a model of imperfect interface with finite strains and damage by asymptotic techniques: An application to masonry structures. *Meccanica* **2018**, *53*, 1645–1660. [CrossRef]
22. Kovtunenko V.A.; Leugering G. A Shape-Topological Control Problem for Nonlinear Crack-Defect Interaction: The Antiplane Variational Model. *SIAM J. Control Optim.* **2016**, *54*, 1329–1351. [CrossRef]
23. Serpilli M.; Lenci S. An overview of different asymptotic models for anisotropic three-layer plates with soft adhesive. *Int. J. Solids Struct.* **2016**, *81*, 130–140. [CrossRef]
24. Geymonat G.; Krasucki F. Analyse asymptotique du comportement en exion de deux plaques collées. *C. R. Acad. Sci. Paris* **1997**, *325*, 307–314.
25. Zaittouni F.; Lebon F.; Licht C. Étude théorique et numérique du comportement d'un assemblage de plaques. *C. R. Mec.* **2002**, *330*, 359–364. [CrossRef]
26. Khludnev A.M.; Kovtunenko V.A. *Analysis of Cracks in Solids*; WIT-Press: Southampton, UK; Boston, MA, USA, 2000.
27. Pelekh B.L. *Theory of Shells with Finite Shear Rigidity*; Naukova Dumka: Kiev, Russia, 1973. (In Russian)
28. Cherdantsev M.; Cherednichenko K. Bending of thin periodic plates. *Calc. Var. Partial Differ. Equ.* **2015**, *54*, 4079–4117. [CrossRef]
29. Lazarev N.P. Problem of equilibrium of the timoshenko plate containing a crack on the boundary of an elastic inclusion with an infinite shear rigidity. *J. Appl. Mech. Tech. Phys.* **2013**, *54*, 322–330. [CrossRef]
30. Rudoy E.M. Asymptotics of the energy functional for a fourth-order mixed boundary value problem in a domain with a cut. *Sib. Math. J.* **2009**, *50*, 341–354. [CrossRef]
31. Rudoy E.M. Invariant integrals for the equilibrium problem for a plate with a crack. *Sib. Math. J.* **2004**, *45*, 388–397. [CrossRef]
32. Rudoy E.M. The Griffith formula and Cherepanov-Rice integral for a plate with a rigid inclusion and a crack. *J. Math. Sci.* **2012**; *186*, 511–529. [CrossRef]
33. Furtsev A.I. Differentiation of energy functional with respect to delamination's length in problem of contact between plate and beam. *Sib. Electron. Math. Rep.* **2018**, *15*, 935–949.
34. Furtsev A.I. A Contact Problem for a Plate and a Beam in Presence of Adhesion. *J. Appl. Ind. Math.* **2019**, *13*, 208–218. [CrossRef]
35. Lazarev N.P.; Rudoy E.M. Optimal size of a rigid thin stiffener reinforcing an elastic plate on the outer edge. *ZAMM Z. Angew. Math. Mech.* **2017**, *97*, 1120–1127. [CrossRef]

Publisher's Note: MDPI stays neutral with regard to jurisdictional claims in published maps and institutional affiliations.

 technologies

MDPI

Article

Effective Complex Properties for Three-Phase Elastic Fiber-Reinforced Composites with Different Unit Cells

Federico J. Sabina [1], Yoanh Espinosa-Almeyda [1], Raúl Guinovart-Díaz [2], Reinaldo Rodríguez-Ramos [2,3,*] and Héctor Camacho-Montes [4]

[1]   Instituto de Investigaciones en Matemáticas Aplicadas y en Sistemas, Universidad Nacional Autónoma de México, Apartado Postal 20-126, Alcaldía Álvaro Obregón, CDMX 01000, Mexico; fjs@mym.iimas.unam.mx (F.J.S.); yoanhealmeyda1209@gmail.com (Y.E.-A.)

[2]   Facultad de Matemática y Computación, Universidad de La Habana, San Lázaro y L, Vedado, Habana 4, La Habana 10400, Cuba; guino@matcom.uh.cu

[3]   Escuela de Ingeniería y Ciencias, Tecnológico de Monterrey, Campus Puebla Atlixcáyotl 5718, Reserva Territorial Atlixcáyotl, Puebla 72453, Mexico

[4]   Instituto de Ingeniería y Tecnología, Universidad Autónoma de Ciudad Juárez, Av. Del Charro 450 Norte Cd. Juárez, Chihuahua 32310, Mexico; hcamacho@uacj.mx

*   Correspondence: reinaldo@matcom.uh.cu

Abstract: The development of micromechanical models to predict the effective properties of multiphase composites is important for the design and optimization of new materials, as well as to improve our understanding about the structure–properties relationship. In this work, the two-scale asymptotic homogenization method (AHM) is implemented to calculate the out-of-plane effective complex-value properties of periodic three-phase elastic fiber-reinforced composites (FRCs) with parallelogram unit cells. Matrix and inclusions materials have complex-valued properties. Closed analytical expressions for the local problems and the out-of-plane shear effective coefficients are given. The solution of the homogenized local problems is found using potential theory. Numerical results are reported and comparisons with data reported in the literature are shown. Good agreements are obtained. In addition, the effects of fiber volume fractions and spatial fiber distribution on the complex effective elastic properties are analyzed. An analysis of the shear effective properties enhancement is also studied for three-phase FRCs.

Keywords: multiphase fiber-reinforced composites; asymptotic homogenization method; effective complex properties; elastic composite

Citation: Sabina, F.J.; Espinosa-Almeyda, Y.; Guinovart-Díaz, R.; Rodríguez-Ramos, R.; Camacho-Montes, H. Effective Complex Properties for Three-Phase Elastic Fiber-Reinforced Composites with Different Unit Cells. *Technologies* 2021, 9, 12. https://doi.org/10.3390/technologies9010012

Received: 4 December 2020
Accepted: 28 January 2021
Published: 1 February 2021

Publisher's Note: MDPI stays neutral with regard to jurisdictional claims in published maps and institutional affiliations.

1. Introduction

Multiphase elastic fiber-reinforced composites (FRCs) are still important in applications because their yields exceed those of their constituents and they offer very interesting properties compared to more conventional materials. Improved levels of functionality can be achieved through the manipulation of physical properties by means of structure optimization [1–3]. Therefore, the effective properties prediction for FRCs by means of micromechanical models and numerical approaches is a useful tool for technological innovation [4–10].

Periodic multiphase elastic FRCs have found applications related to transport problems (conductivity, shear elasticity, dielectric constant, thermal expansion, and others). In this sense, different micromechanical and experimental models have been developed to analyze elastic FRCs. For example, the elastic effective properties of two-phase elastic FRCs with periodic square [11] and hexagonal [12] cells were found by applying the asymptotic homogenization method (AHM). Jiang et al. [13] calculated the effective elastic moduli of FRCs with cylindrical inclusions under longitudinal shear by means of the Eshelby equivalent inclusion concept [14], combining the results of a doubly quasiperiodic Riemann boundary value problem [15]. Artioli et al. [16] applied the AHM to determine the effective longitudinal shear properties of elastic FRCs with radially graded fibers, assuming imperfect interface conditions. Shu and Stanciulescu [17] proposed an analytical approach using multiscale homogenization to characterize FRCs with imperfect interphase through the shear-lag model. Dhimole et al. [18] implemented a multiscale modeling based on homogenization method to predict the accurate mechanical behavior of 3D four-directional braided composites.

In addition, Bisegna and Caselli [19] investigated the effective complex conductivity for a periodic FRC with interfacial impedance and hexagonal symmetry by AHM. Godin computed the complex effective dielectric properties for two-phase FRCs with circular inclusions [20] and for periodic tubular structures [21]. These tubular structures were modeled as a three-phase FRC. Analytic bounds on the complex dielectric constant were reported by the authors of [22]. A correspondence between orthotropic complex-value dielectric media and non-orthotropic elastic media was developed by the authors of [23] through an affine transformation. Mackay and Lakhatia [24] reported the gain and loss enhancement for particulate homogenized composite materials whose active constituents have complex values. Guild et al. [25] analyzed the enhancement of homogenized dielectric effective properties for acoustic waves using multiscale sonic crystals. Luong et al. [26] estimated the complex shear modulus using the least mean square/algebraic helmholtz inversion (LMS/AHI) algorithm for 1D heterogeneous tissue. On the other hand, fiber-matrix interaction region has also been studied for multiphase FRCs. In this case, a thin mesophase is added between the fiber and the matrix in a three-phase FRC. This contact zone is commonly defined as imperfect contact region, see, for instance, [17,27–30].

The AHM has proved to be advantageous in the description of the multiscale mechanics of composite materials. Many studies have dealt with the theoretical bases of the AHM [31–35]. In general, the AHM makes it possible to obtain an effective characterization of the heterogeneous system or phenomenon under study by encoding the information available at the microscale into the so-called effective coefficients. In particular, multiscale AHM take advantage of the information available at the smaller scales of a given heterogeneous medium to predict the effective properties at its larger scales, see, for instance, [36–39]. This, in turn, dramatically reduces the computational complexity of the resulting boundary problems. However, the main disadvantage of AHM is that the analytical solution of the local problems has been derived for only a few composite structures [40,41].

The main aim of this work is the estimation by AHM of the effective elastic complex-values properties for periodic three-phase elastic FRC with complex-valued constituent properties and a parallelogram cell. The out-of-plane case for three-phase composite is considered. Both matrix and fibers have isotropic or transversely isotropic properties, and they are in welded contact. The mathematical statement is presented, and the associated local problems are derived. Simple closed formulas are provided for the shear effective coefficients of three-phase FRCs. Validations of the present model with results reported in literature are shown. The AHM accuracy and convergence is analyzed based on the truncations of the infinite system from the local problems solutions. Also, the effect of volume fraction and spatial fiber distribution on the complex effective elastic properties is analyzed. An example of longitudinal shear enhancement is considered as a function of reinforcement volume fractions for three-phase FRCs with complex-value constituents. Good agreements are obtained.

The novelty of this work is to calculate by AHM the out-of-plane effective properties of periodic three-phase elastic FRC with parallelogram unit cell whose constituents have complex-valued properties. It generalizes earlier works in which the same method has been applied to two- and three-phase FRCs with real-values constituents and square, hexagonal, and parallelogram unit cells, see, for instance, [11,12,40,42]. This generalization allows the study of the shear effective properties enhancement for three-phase FRCs.

2. Statement of the Problem and Method of Solution

2.1. Mathematical Formulation for the Elastic Heterogeneous Media

A heterogeneous periodic three-phase elastic FRC $\Omega \subset \mathbb{R}^3$ (fiber/mesophase/matrix) with a doubly periodic microstructure is studied, which consisted of a parallelepiped array of two concentric and unidirectional cylinders within a homogeneous matrix (Figure 1a). The fibers are infinitely long in the Ox_3−direction and periodically distributed. At the microstructural level, the composite cross-section (periodic unit cell Y) is defined by a matrix with two concentric circles of different radii located at the parallelogram center, see Figure 1c. The constituent elastic properties, belonging to a crystal symmetry class of 6mm, are assumed to be complex numbers. In addition, as a unidirectional FRC, the composite microstructure is considered to remain constant along the reinforcement's direction (i.e., perpendicular to Oy_1y_2−plane of cross-section).

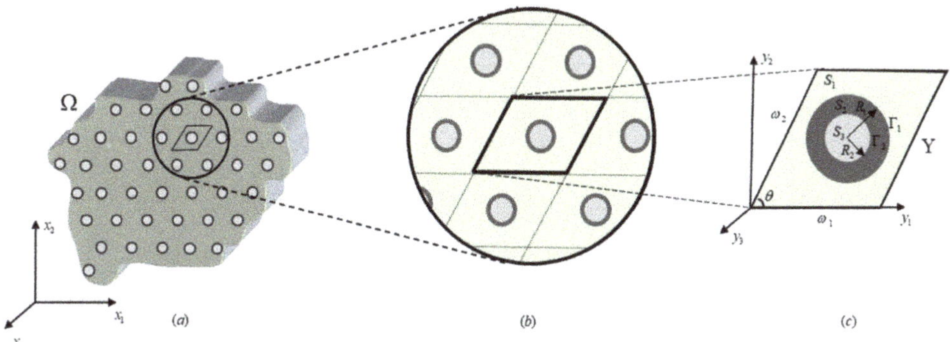

Figure 1. (**a**) Heterogeneous three-phase fiber-reinforced composites (FRC); (**b**) blow-up of periodic structure; (**c**) the representative parallelogram-like unit cell Y.

The periodic parallelogram-like unit cell Y is characterized by a principal angle θ and the periods ω_1 and ω_2 in the Oy_1y_2−plane. Also, it is satisfied that $Y = \overset{3}{\underset{\gamma=1}{\cup}} S_\gamma$ and $S_i \cap S_j = \varnothing, i \neq j$, where S_1 is the region occupied by matrix phase defined by a parallelogram Σ with a central circular hole of radius R_1, contour Γ_1, and volume V_1. S_2 represents the mesophase of volume V_2 and is surrounded by two circular interfaces Γ_1 and Γ_2 of radius R_1 and R_2 $(R_1 > R_2)$, respectively, and S_3 is the central fiber with circular boundary Γ_2 of radius R_2 and volume V_3. The volume V of the cell satisfies that $V = V_1 + V_2 + V_3 = 1$.

In the out-of-plane mechanical deformation state, the mechanical displacement $\mathbf{u} = (u_1, u_2, u_3)$ of a media Ω satisfies that $u_1(x_1, x_2) = u_2(x_1, x_2) = 0, u_3(x_1, x_2) \neq 0$, and the non-null stresses involved in this problem are $\sigma_{13}(x_1, x_2)$ and $\sigma_{23}(x_1, x_2)$. Thus, the static governing equation for an elastic FRC Ω is defined by the partial differential equations system:

$$\frac{\partial}{\partial x_\eta}\left(C_{\eta3\beta3}(\mathbf{x}/\varepsilon)\frac{\partial u_3}{\partial x_\beta}\right) = 0, \text{in } \Omega, \tag{1}$$

where the absence of body forces is considered. Here, $C_{\eta3\beta3}(\mathbf{x}/\varepsilon)$ are Y-periodic and rapidly oscillating coefficients which denotes the elastic stiffness modulus, and η, $\beta = 1, 2$.

The Equation (1) subject to the prescribed boundary conditions:

$$\sigma_{3j}(\mathbf{x})n_j\big|_{\partial\Omega_1} = t_0(\mathbf{x}), \ (j = 1, 2, 3) \text{ on } \partial\Omega_1 \tag{2}$$

$$u_3(\mathbf{x})\big|_{\partial\Omega_2} = g_1(\mathbf{x}), \text{ on } \partial\Omega_2 \tag{3}$$

represent the out-of-plane classical elliptic boundary value problem associated with the linear elasticity theory for heterogeneous structures. Here, $t_0(\mathbf{x})$ is the traction vector on $\partial\Omega_1$, $g_1(\mathbf{x})$ is an infinitely differentiable function on the external boundary $\partial\Omega_2$, and n_j is the component of the outward unit normal vector on $\partial\Omega_1$. The boundary of the composite is partitioned in such a way that $\partial\Omega = \partial\Omega_1 \cup \partial\Omega_2$ and $\partial\Omega_1 \cap \partial\Omega_2 = \varnothing$. In Equations (1)–(3), σ_{3j} and u_3 are the out-of-plane mechanical stresses and displacements.

In addition to Equations (1)–(3), perfect contact conditions at the circular interfaces Γ_s ($s = 1, 2$) are assumed, i.e.,

$$\left[\left[\sigma_{3j}\, n_j\right]\right]\big|_{\Gamma_s} = 0, \quad \left[\left[u_3(\mathbf{x})\right]\right]\big|_{\Gamma_s} = 0, \text{ on } \Gamma_s \tag{4}$$

where $[[f]]\big|_{\Gamma_s} = f^{(s)} - f^{(s+1)}$ means jump of f across Γ_s, see for instance [11].

2.2. Method of Solution: Local Problems and Effective Coefficients

The two-scale AHM [32,33,43] is used to solve the elliptic boundary value problem (Equations (1)–(4)). The solution is found by means of a two-scale asymptotic representation of u_3 in powers of the small geometrical parameter ε by the ansatz:

$$u_3(\mathbf{x}) = u_3^{(0)} + \varepsilon u_3^{(1)}(\mathbf{x},\mathbf{y}) + O(\varepsilon^2), \tag{5}$$

where the macroscale or fast variable "$\mathbf{x}$" and the microscale or slow variable "$\mathbf{y}$" are related by $\mathbf{x} = \varepsilon\mathbf{y}$, and the parameter $\varepsilon = l/L$ is the ratio between the periodic unit cell length (l) and the macroscopic dimension of the composite (L). In Equation (5), the second term $u_3^{(1)}(\mathbf{x},\mathbf{y})$ is a periodic function of $\mathbf{y}$, which represents a correction of the term $u_3^{(0)} \equiv u_3(\mathbf{x})$. Also, it is satisfied that $u_3^{(1)}(\mathbf{x},\mathbf{y}) = {}_{\alpha 3}N(\mathbf{y})\left(\partial u_3^{(0)}\backslash\partial x_\alpha\right)$, where ${}_{\alpha 3}N(\mathbf{y}) \equiv {}_{\alpha 3}N$ is a Y-periodic local function, which is the solution of the local problems. It is possible to obtain an asymptotic solution of the problem (Equations (1)–(4)) when $\varepsilon \to 0$. More details and the rigorous theoretical background of AHM have been described in classical works [31,33,43] and are omitted here.

The out-of-plane local problems on Y, denoted as ${}_{\alpha 3}L$ ($\alpha = 1,2$) for a three-phase elastic FRC (see Figure 1), is stated as follows:

$$\frac{\partial}{\partial y_\beta}\left(C_{1313} + C_{1313}\frac{\partial_{\alpha 3}N}{\partial y_\beta}\right) = 0, \text{ in } S_\gamma\,(\gamma = 1,\, 2,\, 3), \tag{6}$$

$${}_{\alpha 3}N^{(s)}\Big|_{\Gamma_s} = {}_{\alpha 3}N^{(s+1)}\Big|_{\Gamma_s}, \text{ on } \Gamma_s \tag{7}$$

$$\left[\left(C_{1313}^{(s)}\frac{\partial_{\alpha 3}N^{(s)}}{y_\beta} - C_{1313}^{(s+1)}\frac{\partial_{\alpha 3}N^{(s+1)}}{y_\beta}\right)n_\beta\right]\Big|_{\Gamma_s} = -\left[\left(C_{1313}^{(s)} - C_{1313}^{(s+1)}\right)(\delta_{1\alpha}\,n_1 + \delta_{2\alpha}\,n_2)\right]\Big|_{\Gamma_s}, \text{ on } \Gamma_s, \tag{8}$$

where $\delta_{1\alpha}$ and $\delta_{2\alpha}$ are the Kronecker's delta functions related to the ${}_{13}L$ and ${}_{23}L$ local problems, respectively. To guarantee the uniqueness of the local problems solutions, the local functions should satisfy the null average condition $\langle {}_{\alpha 3}N\rangle = 0$ on Y, where $\langle F\rangle = (1/|Y|)\int_Y F(\mathbf{y})\,d\mathbf{y}$ and $|Y|$ is the area of Y.

Once the ${}_{\alpha 3}L$ out-of-plane local problems (Equations (6)–(8)) are solved, the corresponding effective elastic coefficients can be calculated by the formulas:

$$\begin{aligned} C_{1313}^* &= \langle C_{1313}(\mathbf{y}) + C_{1313}(\mathbf{y})_{13}N_{,1}(\mathbf{y})\rangle, \\ C_{2313}^* &= \langle C_{1313}(\mathbf{y})_{13}N_{,2}(\mathbf{y})\rangle, \end{aligned} \quad \text{associate with } {}_{13}L \text{ local problem,} \tag{9}$$

$$\begin{aligned} C_{1323}^* &= \langle C_{1313}(\mathbf{y})_{23}N_{,1}(\mathbf{y})\rangle, \\ C_{2323}^* &= \langle C_{1313}(\mathbf{y}) + C_{1313}(\mathbf{y})_{23}N_{,2}(\mathbf{y})\rangle, \end{aligned} \quad \text{associate with } {}_{23}L \text{ local problem.} \tag{10}$$

Notice that the out-of-plane effective elastic coefficients (Equations (9) and (10)) depend on the local functions ${}_{13}N(\mathbf{y})$ and ${}_{23}N(\mathbf{y})$ relative to the ${}_{13}L$ and ${}_{23}L$ local problems, respectively. Then, $_{13}N(\mathbf{y})$ and $_{23}N(\mathbf{y})$ need to be found. Therefore, an analytical solution of Equations (6)–(8) is determined.

On the other hand, the homogenized elastic problem equivalent to the boundary value problem (Equations (1)–(3)) is defined by the equation system

$$C^{*}_{3\alpha3\beta}\frac{\partial^2 u_3^{(0)}}{\partial x_\alpha \partial x_\beta} = 0, \ (\alpha, \beta = 1, 2) \text{ on } \overline{\Omega}, \tag{11}$$

subject to the homogenized boundary conditions

$$u_3^{(0)}\Big|_{\partial\overline{\Omega}_u} = \overline{g}_1(\mathbf{x}), \quad \sigma_{3j}^{(0)} n_j\Big|_{\partial\overline{\Omega}_\sigma} = \overline{t}_0, \text{ on } \partial\overline{\Omega} = \partial\overline{\Omega}_u \cup \partial\overline{\Omega}_\sigma \tag{12}$$

where $u_3^{(0)}$ is the solution of the homogenized problem, and $C^{*}_{3\alpha3\beta}$ are the out-of-plane effective elastic coefficients defined in Equations (9) and (10).

2.3. Analytical Solution of the Local Problems

By means the potential methods of a complex variable theory, the solution $_{\alpha3}N$ of the $_{\alpha3}L$ local problem (Equations (6)–(8)) is calculated. Here, the doubly periodic Weierstrass' elliptic functions are used to obtain an analytical solution, i.e., the double periodic solution $_{\alpha3}N$ is found in terms of Laurent and powers expansions as a function of $z = y_1 + iy_2$, see, for instance, [44,45], as follows

$$_{\alpha3}N^{(1)} = \text{Re}\left\{ _{\alpha3}a_0 z\, R_1^{-1} + \sum_{p=1}^{\infty} {}^{o}\sum_{k=1}^{\infty} {}^{o}\,_{\alpha3}a_k \eta_{kp} R_1^{-p} z^p + \sum_{p=1}^{\infty} {}^{o}\,_{\alpha3}a_p\, R_1^p z^{-p} \right\}, \text{ at matrix region } S_1, \tag{13}$$

where $\eta_{kp} = -\frac{(k+p-1)!}{p!(k-1)!} R^{k+p} S_{k+p}$ with $S_{k+p} = \sum_{m,n} \beta_{mn}^{-(k+p)} = \sum_{m,n}(mw_1 + nw_2)^{-(k+p)}, m^2 + n^2 \neq 0, k+p > 2$, and $k, p = 1, 3, 5, \cdots$. By

$$_{\alpha3}N^{(2)} = \text{Re}\left\{ \sum_{p=1}^{\infty} {}^{o}\,_{\alpha3}b_p\, R_2^{-p} z^p + \sum_{p=1}^{\infty} {}^{o}\,_{\alpha3}b_{-p}\, R_1^p\, z^{-p} \right\}, \text{ at mesophase region } S_2, \tag{14}$$

and

$$_{\alpha3}N^{(3)} = \text{Re}\left\{ \sum_{p=1}^{\infty} {}^{o}\,_{\alpha3}c_p\, R_2^{-p}\, z^p \right\}, \text{ at fiber region } S_3. \tag{15}$$

In Equations (13)–(15), the coefficients $_{\alpha3}a_0$, $_{\alpha3}a_p$, $_{\alpha3}b_p$, $_{\alpha3}b_{-p}$, and $_{\alpha3}c_p$ are complex and undetermined numbers. They need to be determined in order to find the $_{\alpha3}L$ local problems solution and the out-of-plane effective elastic coefficients (Equations (9) and (10)). Here, it can be highlighted that the summation symbol with superscript $\sum^{o}$ means that the sum only runs over odd integers, and the symbols Re and Im represent the real and imaginary parts of complex numbers, respectively. Details of Laurent and powers expansions and its relationship with the double periodic elliptic Weierstrass function $\wp(w_1, w_2; z)$ of periods w_1 and w_2, and related expressions can be found in [45,46].

From the double periodicity condition of $_{\alpha3}N$, it is satisfied that:

$$_{\alpha3}N(z + w_\alpha) - {}_{\alpha3}N(z) = \text{Re}\left\{ _{\alpha3}a_0 R_1^{-1} w_\alpha + {}_{\alpha3}a_1 \delta_\alpha R_1 \right\}, \tag{16}$$

see, for instance, [45]. Then, it can be proved that $_{\alpha3}a_0$ is linked with $_{\alpha3}a_1$ by the equation:

$$_{\alpha3}a_0 = -R_1^2 H_1\,_{\alpha3}\overline{a}_1 - R_1^2 H_2\,_{\alpha3}a_1, \tag{17}$$

where $H_1 = \left(\overline{\delta}_1\overline{w}_2 - \overline{\delta}_2\overline{w}_1\right)/(w_1\overline{w}_2 - w_2\overline{w}_1)$, $H_2 = \left(\delta_1\overline{w}_2 - \delta_2\overline{w}_1\right)/(w_1\overline{w}_2 - w_2\overline{w}_1)$ and $_{\alpha3}\overline{a}_1$ is the complex conjugate of $_{\alpha3}a_1$. In addition, $\delta_\alpha = 2\zeta(w_\alpha/2)$ is the quasiperiodic condition and $\zeta(z)$ is the Weierstrass quasiperiodic Zeta function defined by $\zeta(z) = z^{-1} + \sum'_{m,n} \left[(z - \beta_{mn})^{-1} + \beta_{mn}^{-1} + z\beta_{mn}^{-2} \right]$ where $\sum'_{m,n}$ means that the summation does not include the point $(0, 0)$, see [45].

Finally, replacing Equations (13)–(15) into the interface conditions (Equations (7) and (8)) and after some mathematical manipulations, we obtain the normal infinity system of linear equations [47] with unknown complex constants $_{\alpha 3}a_p$, in compact form:

$$_{\alpha 3}\bar{a}_p + \chi_1 R_1^2 H_1 \delta_{1p\,\alpha 3}\bar{a}_1 + \chi_1 R_1^2 H_2 \delta_{1p\,\alpha 3}a_1 + \chi_p \sum_{k=1}^{\infty} {}^* {}_{\alpha 3}a_k W_{kp} = ER_1 \delta_{1p}[\delta_{1\alpha} - i\delta_{2\alpha}], \quad (18)$$

where $\chi_p = \dfrac{(V_2+V_3)^p (k_1+k_2)(1-k_1) + V_3^p (k_1-k_2)(1+k_1)}{(V_2+V_3)^p (k_1+k_2)(1+k_1) + V_3^p (k_1-k_2)(1-k_1)}$, $E = \chi_1$, and $k_s = C_{1313}^{(s+1)}/C_{1313}^{(1)}$ $(s = 1, 2)$. In Equation (18), $W_{kp} = \sqrt{pk^{-1}}\,\eta_{kp}$, V_2 and V_3 are the volume fractions of the mesophase and the central fiber, and the symbol δ_{1p} is the Kronecker's delta function. Also, the system solution of each local problem $_{\alpha 3}L$ $(\alpha = 1, 2)$ depends on the elastic constituent properties, the phase volume fractions and the fibers spatial distribution within the matrix. Details of the system construction can be found in [40] and are omitted here. Once the unknown constants $_{\alpha 3}a_p (p = 1, 3, 5, \cdots)$ are calculated, the local problem solution and the effective coefficients can be determined. Details on the system solution is given in Appendix A.

The non-null out-of-plane effective properties for three-phase elastic FRC are listed as follows:

$$C_{1313}^* - iC_{2313}^* = \langle C_{1313} \rangle - [[C_{1313}]]_2 \frac{V_2+V_3}{\chi_1 R_1}[(\chi_1 + 1)_{13}\bar{a}_1 - R_1 E] - [[C_{1313}]]_3 \frac{V_3}{R_2^2}\,_{13}c_1, \quad (19)$$

$$C_{1323}^* - iC_{2323}^* = -i\langle C_{1313} \rangle - [[C_{1313}]]_2 \frac{V_2+V_3}{\chi_1 R_1}[(\chi_1 + 1)_{23}\bar{a}_1 + iR_1 E] - [[C_{1313}]]_3 \frac{V_3}{R_2^2}\,_{23}c_1, \quad (20)$$

where $\langle C_{1313} \rangle = C_{1313}^{(1)}V_1 + C_{1313}^{(2)}V_2 + C_{1313}^{(3)}V_3$ is the Voigt's average, $A_{21} = \left[R_1^2(k_1 + k_2) + R_2^2(k_1 - k_2) \right]/2R_1 R_2$, and the constant is defined by

$$_{\alpha 3}c_1 = (k_1(\chi_1 + 1)/\chi_1 A_{21})_{\alpha 3}\bar{a}_1 - \left\{ \left[2k_1^2 R_1^2 - k_1(k_1 - k_2)(R_1^2 - R_2^2) \right]/(2A_{21}k_1 R_1) \right\}[\delta_{1\alpha} - i\delta_{2\alpha}]. \quad (21)$$

Simple closed-form formulas for the effective properties equivalent to Equations (19) and (20) are given in Appendix B.

3. Numerical Results

The accuracy of the AHM model is determined through comparisons with other results reported in the literature for three-phase elastic FRC (fiber/mesophase/matrix) with complex-values constituent properties. In addition, the effect of the volume fraction of inclusions and the spatial fiber distribution on the complex effective elastic properties is investigated. Also, an example of shear enhancement is reported as a function of reinforcement volume fractions.

Limit cases for the present model can be determined when Equations (19) and (20) and system (18) are reduced to those that represent a two- and three-phase FRC with parallelogram-like unit cell. In these cases, isotropic or transversely isotropic constituents are considered with real-values elastic properties, as reported by the authors of [11,40,42].

The real and imaginary parts of the complex effective elastic properties are given for some biological applications, for example, the behaviors of biological tissues, skeletal muscle, sclera, and other ones in which material properties depends on time [48–51]. This behavior can be analyzed in viscoelastic materials in which the shear wave speed is connected to the enhancement and lessening of the shear modulus. Besides, in the context of transport properties, the mathematical statement for shear linear responses is identical to conductivity, dielectric permittivity, and so on in equivalent media [52].

In the literature, to the authors' knowledge, the longitudinal shear homogenization problem has not been reported with complex-values coefficients. A comparison with

Godin [21] is possible because the governing equation for both models has the same mathematical formula, although they model different physical magnitudes. Both models find an infinity equation system to solve the problem. However, there are differences between the model implementations. The Godin model goes directly to a representative element of analysis and the governing equation is directly solved. The effective properties are proposed as a linear relation between averaged physical magnitudes. The AHM described herein in previous sections is based on a procedure that ends up in the solutions of local problems, the boundary conditions, and the effective properties.

In Table 1, the real (Re) and imaginary (Im) parts of complex effective property C^*_{1313} are illustrated for two different three-phase FRCs as a function of normalized radius $h = R_1/l$ with a square and hexagonal unit cell, respectively. Here, $h \leq 1/2$, R_1 is the outer interface radius, and l represents the minimum distance between the centers of the fibers. In addition, an analysis of the relative error is shown: Error $= [(\text{AHM} - \text{Godin})/\text{AHM}] \times 100\%$ [21]. Hence, the numerical values of the effective property C^*_{1313} is compared with the complex effective dielectric constant ε^* provided by the authors of [21]. The numerical calculations are carried out considering that the matrix, the mesophase, and the fiber have the complex-values properties $C^{(1)}_{1313} = 5 - 4i$, $C^{(2)}_{1313} = 80 - 2i$, and $C^{(3)}_{1313} = 2 - 4i$ for the FRC with the square array and $C^{(1)}_{1313} = 1$, $C^{(2)}_{1313} = 8 - 40i$, and $C^{(3)}_{1313} = 2 - 4i$ for the FRC with the hexagonal array. In both three-phase FRCs, the concentric fibers radius relation is $R_2^2/R_1^2 = 0.81$. As can be seen, good agreements are achieved by AHM for only a system truncation (Equation (18)) to a finite order $N_0 = 2$. Only a very slight discrepancy is observed near the close-packing condition, although the relative error is below 0.27% in every case.

Table 1. Real and imaginary parts of the complex effective property C^*_{1313} as a function of normalized radius for two three-phase FRCs with a square and hexagonal unit cell, respectively.

| | C^*_{1313} (GPa) (Square Unit Cell, $\theta = 90°$) | | | | | |
| | Re (C^*_{1313}) | | | Im (C^*_{1313}) | | |
h	AHM	Godin [21]	Error (%)	AHM	Godin [21]	Error (%)
0.1	5.12292	5.12291	0	−4.02626	−4.02626	0
0.2	5.50633	5.50633	0	−4.10103	−4.10103	0
0.3	6.19810	6.19810	0	−4.20863	−4.20863	0
0.4	7.29680	7.29682	0.0003	−4.30363	−4.30364	0.0007
0.499	9.00105	8.99915	0.0568	−4.21080	−4.22202	0.2663

| | C^*_{1313} (GPa) (Hexagonal Unit Cell, $\theta = 60°$) | | | | | |
| | Re (C^*_{1313}) | | | Im (C^*_{1313}) | | |
h	AHM	Godin [21]	Error (%)	AHM	Godin [21]	Error (%)
0.1	1.06782	1.06782	0	−0.01569	−0.01569	0
0.2	1.29996	1.29996	0	−0.07777	−0.07777	0
0.3	1.81491	1.81491	0	−0.26416	−0.26416	0
0.4	2.97891	2.97891	0	−0.99392	0.99393	0
0.499	3.85417	3.85517	0.0261	−6.27906	−6.27919	0.0021

In Table 2, the variations of the real and imaginary parts of the complex effective shears C^*_{1313}, C^*_{1323}, and C^*_{2323} in terms of system truncate orders N_0 ($N_0 = 1, 2, 3, 5, 7, 9, 11$) are presented for a three-phase elastic FRC with a parallelogram unit cell of $\theta = 75°$. In addition, two different normalized radii $h = 0.4$ and $h = 0.499$ (close to percolation value) are also considered. For the analysis, the material properties of matrix, mesophase, and fiber have isotropic complex properties, such as $C^{(1)}_{1313} = 5 - 4i$, $C^{(2)}_{1313} = 80 - 2i$, and

$C^{(3)}_{1313} = 2 - 4i$. The relation between the concentric fiber's radius is $R_2^2 / R_1^2 = 0.81$. Notice that the AHM convergence is achieved quickly when low values of $h \leq 0.4$ are assumed, i.e., only smaller values of $N_0 \leq 3$ are needed. Therefore, truncations of higher order N_0 must be considered for high values of h, as well as for higher contrast among the matrix mesophase and fiber properties, in order to obtain a better accuracy. For example, $N_0 \geq 9$ is required to obtain the effective properties values with at least five accuracy digits when $h = 0.499$. Similar conclusions are achieved when the same analysis is developed for a three-phase elastic FRC with a parallelogram unit cell of $\theta = 75°$ and complex-values constituent properties $C^{(1)}_{1313} = 1$, $C^{(2)}_{1313} = 8 - 40i$, and $C^{(3)}_{1313} = 2 - 4i$.

Table 2. Real and imaginary parts of the complex effective shears C^{*}_{1313}, C^{*}_{1323}, and C^{*}_{2323} obtained by AHM in term of order system N_0 for a three-phase FRC with a parallelogram unit cell of $\theta = 75°$ and for two different normalized radii $h = 0.4$ and $h = 0.499$.

N_0	Effective complex shears C^{*}_{1313}, C^{*}_{1323} and C^{*}_{2323} (in GPa) for a three-phase FRC with parallelogram unit cell of $\theta = 75°$ and $h = 0.4$.					
	Re (C^{*}_{1313})	Im (C^{*}_{1313})	Re (C^{*}_{1323})	Im (C^{*}_{1323})	Re (C^{*}_{2323})	Im (C^{*}_{2323})
1	7.394470	−4.308182	0.022871	0.0098056	7.382213	−4.313437
2	7.395010	−4.307804	0.022972	0.0099212	7.382699	−4.31312
3	7.395020	−4.307797	0.022966	0.0099148	7.382713	−4.31311
5	7.395021	−4.307796	0.022966	0.0099150	7.382714	−4.313109
7	7.395021	−4.307796	0.022966	0.0099150	7.382714	−4.313109
9	7.395021	−4.307796	0.022966	0.0099150	7.382714	−4.313109
11	7.395021	−4.307796	0.022966	0.0099150	7.382714	−4.313109

N_0	Effective complex shears C^{*}_{1313}, C^{*}_{1323}, and C^{*}_{2323} (in GPa) for a three-phase FRC with a parallelogram unit cell of $\theta = 75°$ and $h = 0.499$.					
	Re (C^{*}_{1313})	Im (C^{*}_{1313})	Re (C^{*}_{1323})	Im (C^{*}_{1323})	Re (C^{*}_{2323})	Im (C^{*}_{2323})
1	9.156983	−4.231357	0.052276	0.016613	9.128968	−4.240260
2	9.174364	−4.209425	0.063003	0.033689	9.140601	−4.227480
3	9.175411	−4.205838	0.061539	0.029078	9.142432	−4.221421
5	9.176241	−4.201475	0.062230	0.031106	9.142892	−4.218145
7	9.176080	−4.200378	0.062254	0.031390	9.142718	−4.21720
9	9.175964	−4.200090	0.062271	0.031380	9.142593	−4.216907
11	9.175904	−4.199996	0.062285	0.031371	9.142526	−4.216808

Table 3 shows the real and imaginary parts of the overall out-of-plane shear properties C^{*}_{1313}, C^{*}_{1323}, and C^{*}_{2323} for two three-phase FRCs with different parallelogram-like unit cells. The numerical values are computed considering four different parallelogram-like unit cells (i.e., parallelogram cells characterized by a principal angle θ equal to $45°$, $60°$, $75°$, and $90°$), a system order truncation $N_0 = 10$, and $h = 0.38$ (value of the normalized radius near to the percolation point volume 0.38268, for $45°$). In addition, the composite structure-property relationship is also analyzed. From Table 3, it is noticed that, when the periodic unit cells are characterized by parallelograms with θ different to $(60°)$ and $(90°)$, the composites belong to monoclinic symmetric class, i.e., 13 non-null effective elastic constants are attained. However, in the out-of-plane case, only $C^{*}_{1313} \neq C^{*}_{2323}$ and $C^{*}_{1323} = C^{*}_{2313} \neq 0$ are remained. In the case of the periodic hexagonal $(60°)$ and square $(90°)$ unit cells, the composite behavior is transversely isotropic, i.e., $C^{*}_{1313} = C^{*}_{2323}$ and $C^{*}_{1323} = C^{*}_{2313} = 0$. These results have also been satisfied in elastic FRCs with real effective properties, see, for instance, [40]. In addition, it can be concluded that a decrease (increase) in the real (imaginary) part of the complex effective shears C^{*}_{1313} and C^{*}_{2323} resulted as the angle of the periodic unit cell increased. Higher values for the real and imaginary parts of C^{*}_{1313} and C^{*}_{2323} are obtained when $\theta = 30°$ and the normalized radius is the same. Besides, for the composite with complex-value constituents $C_{1313}^{(1)} = 5 - 4i$, $C_{1313}^{(2)} = 80 - 2i$, and $C_{1313}^{(3)} = 2 - 4i$, the real and imaginary parts of C_{1323}^{*} are negative for $30° \leq \theta < 60°$ and positive for $60° < \theta < 90°$. In a composite with complex-value constituents $C_{1313}^{(1)} = 1$, $C_{1313}^{(2)} = 8 - 40i$, and $C_{1313}^{(3)} = 2 - 4i$, the real (imaginary) part of C_{1323}^{*} run from negative (positive) to positive (negative) when $30° \leq \theta < 90°$. It should be noted that there are three different effective behaviors.

Table 3. Real and imaginary parts of the complex effective shears C_{1313}^{*}, C_{1323}^{*}, and C_{2323}^{*} for two three-phase FRCs with different parallelogram-like unit cells and a normalized radius $h = 0.38$.

θ	Real and imaginary parts of the effective complex shears C_{1313}^{*}, C_{1323}^{*} and C_{2323}^{*} (in GPa) of a three-phase FRC with constituent properties $C_{1313}^{(1)} = 5 - 4i$, $C_{1313}^{(2)} = 80 - 2i$, and $C_{1313}^{(3)} = 2 - 4i$.					
	Re (C_{1313}^{*})	Im (C_{1313}^{*})	Re (C_{1323}^{*})	Im (C_{1323}^{*})	Re (C_{2323}^{*})	Im (C_{2323}^{*})
45°	7.959917	−4.365499	−0.137684	−0.079617	8.235285	−4.206265
60°	7.402436	−4.315333	0	0	7.402436	−4.315333
75°	7.123041	−4.295037	0.018346	0.007566	7.113210	−4.299092
90°	7.036910	−4.290163	0	0	7.036910	−4.290163

N_0	Real and imaginary parts of the effective complex shears C_{1313}^{*}, C_{1323}^{*}, and C_{2323}^{*} (in GPa) of a three-phase FRC with constituent properties $C_{1313}^{(1)} = 1$, $C_{1313}^{(2)} = 8 - 40i$, and $C_{1313}^{(3)} = 2 - 4i$.					
	Re (C_{1313}^{*})	Im (C_{1313}^{*})	Re (C_{1323}^{*})	Im (C_{1323}^{*})	Re (C_{2323}^{*})	Im (C_{2323}^{*})
45°	2.947561	−1.280780	−0.001267	1.090406	2.950095	−3.461592
60°	2.659276	−0.741382	0	0	2.659276	−0.741382
75°	2.412547	−0.589351	0.044877	−0.044002	2.388498	−0.565770
90°	2.332192	−0.538175	0	0	2.332192	−0.538175

In Figure 2, an analysis of enhancement of the real part of the shear effective property C_{1313}^{*} is illustrated for a three-phase FRC with periodic hexagonal unit cell. Notice that, for this type of unit cell, $C_{2323}^{*} = C_{1313}^{*}$ and $C_{1323}^{*} = C_{2313}^{*} = 0$. Here, the enhancement is studied as function of reduced mesophase filling fraction V_2, i.e., the enhancement of C_{1313}^{*} i obtained for four configurations of the annular inclusion (mesophase and fiber inclusions) with fixed volume fraction $V_2 + V_3$. For this analysis, two different three-phase FRCs are considered with complex-values constituents $C_{1313}^{(1)} = 1 - 1i$, $C_{1313}^{(3)} = 1.02 - 1i$, and $C_{1313}^{(2)} = 0.99 - 0.5i$ (Figure 2a) or $C_{1313}^{(2)} = 1.01 - 0.5i$ (Figure 2b). These properties are considered assuming that the mesophase properties attain an interval of realistic physical properties as a combination of the matrix and fiber phases. From Figure 2, it is important to note that the real part of C_{1313}^{*} increase as $V_2 + V_3$ increase when $0 < V_3 < 1$, and that higher values of C_{1313}^{*} are always obtained in comparison when $V_3 = 0$ (two-phase FRC—solid black line). In the figures, the red dashed dot and the blue dashed lines represent the real part values of the matrix and mesophase properties, respectively. In addition, it can be concluded that the imaginary part values of both three-phase FRCs vary monotonically between the values of the matrix and mesophase phases. This picture, in which the three-phase FRCs are considered with distinct constituents, is different than Figure 2 reported by the authors of [7]. In this reference, the enhancement was analyzed for a three-phase FRC with the same values of matrix and fiber and different mesophase, as an annular inclusion. Besides, it should be noted that the volume fraction interval where enhancement appeared is much larger. In one case, it is the whole interval.

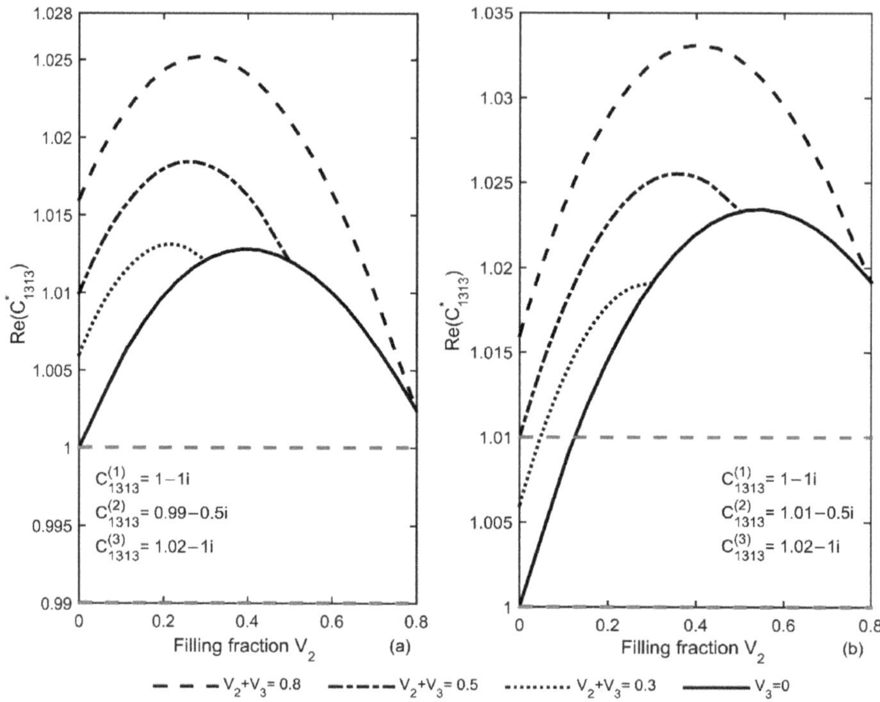

Figure 2. Enhancement of the real part of effective shear property C^{*}_{1313} as a function of reduced mesophase filling fraction V_2 for two three-phase FRCS with hexagonal unit cell and mesophase complex-values property (**a**) $C^{(2)}_{1313} = 0.99 - 0.5i$ and (**b**) $C^{(2)}_{1313} = 1.01 - 0.5i$.

4. Conclusions

In this work, the effective shear properties of periodic three-phase fiber-reinforced composites with complex-values constituent properties and parallelogram unit cells are calculated by AHM. Easy-to-handle formulas and fast numerical implementation are derived for all shear effective properties. We conclude that:

(i) The fiber spatial distribution, represented as parallelogram-like unit cell, is capable of describing three class of symmetry point group: tetragonal 4 mm (square unit cell), hexagonal 6 mm (hexagonal unit cell), and monoclinic 2 (other parallelogram unit cells) structures.

(ii) The enhancement in the shear effective property C^{*}_{1313} is more remarkable for three-phase FRC than two-phase FRC.

(iii) The volume fraction interval where enhancement appeared is larger for a three-phase FRC than the interval for the two-phase FRC.

(iv) The presence of negative values for the real and imaginary parts of C^{*}_{1323} appears for some parallelogram unit cells.

(v) The manipulation of the mesophase can be used as a way to enhance the real and imaginary parts of the shear elastic properties.

(vi) The numerical results prove that the AHM is an accurate and efficient approach for the study of FRC with a mesophase and for different spatial fiber distributions in a matrix.

Author Contributions: Conceptualization, F.J.S., R.R.-R.; methodology, Y.E.-A., H.C.-M.; software and validation, R.G.-D., Y.E.-A.; formal analysis and investigation, F.J.S., R.R.-R.; writing—original draft preparation, Y.E.-A., R.R.-R.; writing—review and editing, H.C.-M., Y.E.-A.; supervision, F.J.S., R.R.-R. All authors have read and agreed to the published version of the manuscript.

Funding: This research received no external funding.

Institutional Review Board Statement: Not applicable.

Informed Consent Statement: Not applicable.

Acknowledgments: Y.E.-A. gratefully acknowledges the Program of Postdoctoral Scholarships of DGAPA from UNAM, Mexico. F.J.S. thanks the funding of PAPIIT-DGAPA-UNAM IA100919. R.R.-R. would like to thank the Department of Mathematics and Mechanics at IIMAS, UNAM. H.C.-M. and Y.E.-A. are grateful for the support of the CONACYT Basic science grant A1-S-9232.

Conflicts of Interest: The authors declare no conflict of interest.

Appendix A

The infinite system (Equation (18)) is solved by truncation to a finite order N_0 through 4×4 blocks for different values of k and p (odd natural numbers) with the unknown complex coefficients $_{\alpha 3}a_p$. The sub-matrix systems (4×4 blocks) are solved by the Gauss's method. A fast convergence of successive truncations is assured due to the system regularity, hence, the method of successive approximations can be applied, see for instance [47].

The above infinite system (Equation (18)) can be rewritten in matrixial form as follows:

$$\left[I + \chi_1 R_1^2 J \delta_{1p} + W \right] X = R_1 E \delta_{1p} B, \tag{A1}$$

where I is the unit matrix, $J = \begin{pmatrix} h_{11} + h_{12} & h_{21} - h_{22} \\ -h_{21} - h_{22} & h_{11} - h_{12} \end{pmatrix}$, $W \equiv W(w_{kp}) = \chi_p \begin{pmatrix} w_{1kp} & -w_{2kp} \\ -w_{2kp} & -w_{1kp} \end{pmatrix}$ is made up of different blocks of order 2 and the infinite vectors X and B are defined by $X = (x_1, y_1, x_3, y_3, \ldots)^T$ and $B = (\delta_{1\alpha}, \delta_{2\alpha})^T$, respectively.

In order to find the solution of system (A1), it is reduced by means of two separate systems of real and imaginary magnitudes, considering that $_{\alpha 3}a_k = {}_{\alpha 3}x_k + i_{\alpha 3}y_k$, $W_{kp} = w_{1kp} + iw_{2kp}$ and $H_\alpha = h_{1\alpha} + ih_{2\alpha}$ where $_{\alpha 3}x_k, {}_{\alpha 3}y_k, w_{1kp}, w_{2kp}, h_{1\alpha}$ and $h_{2\alpha}$ are real numbers and $i^2 = -1$, see for instance [40].

Consequently, following the same procedure applied in examples of alike systems [7,40,53], the solution of Equation (18) can be computed in the matrixial form by

$$X = R_1 E \left[I + \chi_1 R_1^2 J - \chi_1 N_1 \left(I + W \right)^{-1} N_2 \right]^{-1} B, \tag{A2}$$

where $N_1 = \chi_1 \begin{pmatrix} w_{1k1} & -w_{2k1} \\ -w_{2k1} & -w_{1k1} \end{pmatrix}$ and $N_2 = \chi_p \begin{pmatrix} w_{11p} & -w_{21p} \\ -w_{21p} & -w_{11p} \end{pmatrix}$ are infinite matrices of 2×2 blocks of by rows and by columns, respectively. Here, $k = 2t + 1$, $p = 2t_1 + 1$, and the usual index sum is applied by t, $t_1 = 1, 2, 3, \cdots$.

Therefore, the system solution $_{\alpha 3}a_1$ associated to the local problem $_{\alpha 3}L$ $(\alpha = 1, 2)$ is explicitly determined as follows:

$$_{13}a_1 = R_1 E (\ 1 \quad i \ ) Z^{-1} \begin{pmatrix} 1 \\ 0 \end{pmatrix} = \frac{R_1 E(z_{22} - iz_{21})}{z_{11}z_{22} - z_{12}z_{21}}, \tag{A3}$$

$$_{23}a_1 = R_1 E (\ 1 \quad i \ ) Z^{-1} \begin{pmatrix} 0 \\ 1 \end{pmatrix} = -\frac{R_1 E(z_{12} - iz_{11})}{z_{11}z_{22} - z_{12}z_{21}}, \tag{A4}$$

where the matrix $Z \equiv \begin{pmatrix} z_{11} & z_{12} \\ z_{21} & z_{22} \end{pmatrix} = \left[I + \chi_1 R_1^2 J - \chi_1 N_1 \left(I + W \right)^{-1} N_2 \right]$ and Z^{-1} is the inverse matrix of Z.

Appendix B

An equivalent representation of the effective coefficients Equations (19) and (20) can be obtained replacing Equations (21), (A3) and (A4) into Equations (19) and (20), such as:

$$C^*_{1313} = \langle C_{1313} \rangle - C^{(1)}_{1313} B_1 [(\chi_1 + 1)z_{22} - |Z|] - C^{(1)}_{1313} C_1, \tag{A5}$$

$$C^*_{2313} = C^{(1)}_{1313} B_1 (\chi_1 + 1)z_{21}, \tag{A6}$$

$$C^*_{1323} = C^{(1)}_{1313} B_1 (\chi_1 + 1)z_{12}, \tag{A7}$$

$$C^*_{2323} = \langle C_{1313} \rangle - C^{(1)}_{1313} B_1 [(\chi_1 + 1)z_{11} - |Z|] - C^{(1)}_{1313} C_1 \tag{A8}$$

where $B_1 = \dfrac{(V_2+V_3)[(k_1+k_2)(1-k_1)V_2+2k_1(1-k_2)V_3]^2}{\chi_1|Z|[(k_1+k_2)(1+k_1)V_2+2k_1(1+k_2)V_3][(k_1+k_2)V_2+2k_1V_3]}$ and $C_1 = \dfrac{(k_1-k_2)^2 V_2 V_3}{(k_1+k_2)V_2+2k_1V_3}$.

References

1. Bonfoh, N.; Coulibaly, M.; Sabar, H. Effective properties of elastic composite materials with multi-coated reinforcements: A new micromechanical modelling and applications. *Compos. Struct.* **2014**, *115*, 111–119. [CrossRef]
2. Prashanth, S.; Subbaya, K.M.; Nithin, K.; Sachhidananda, S. Fiber Reinforced Composites—A Review. *J. Mater. Sci. Eng.* **2017**, *6*, 1000341.
3. Egbo, M.K. A fundamental review on composite materials and some of their applications in biomedical engineering. *J. King Saud. Univ. Eng. Sci.* **2020**, in press. [CrossRef]
4. Wang, M.; Pan, N. Predictions of effective physical properties of complex multiphase materials. *Mater. Sci. Eng. R Reports* **2008**, *63*, 1–30. [CrossRef]
5. Beicha, D.; Kanit, T.; Brunet, Y.; Imad, A.; El Moumen, A.; Khelfaoui, Y. Effective transverse elastic properties of unidirectional fiber reinforced composites. *Mech. Mater.* **2016**, *102*, 47–53. [CrossRef]
6. Dinzart, F.; Sabar, H.; Berbenni, S. Homogenization of multi-phase composites based on a revisited formulation of the multi-coated inclusion problem. *Int. J. Eng. Sci.* **2016**, *100*, 136–151. [CrossRef]
7. Sabina, F.J.; Guinovart-Díaz, R.; Espinosa-Almeyda, Y.; Rodríguez-Ramos, R.; Bravo-Castillero, J.; López-Realpozo, J.C.C.; Guinovart-Sanjuán, D.; Böhlke, T.; Sánchez-Dehesa, J. Effective transport properties for periodic multiphase fiber-reinforced composites with complex constituents and parallelogram unit cells. *Int. J. Solids Struct.* **2020**, *204–205*, 96–113. [CrossRef]
8. Nasirov, A.; Fidan, I. Prediction of mechanical properties of fused filament fabricated structures via asymptotic homogenization. *Mech. Mater.* **2020**, *145*, 103372. [CrossRef]
9. Nasirov, A.; Gupta, A.; Hasanov, S.; Fidan, I. Three-scale asymptotic homogenization of short fiber reinforced additively manufactured polymer composites. *Compos. Part B Eng.* **2020**, *202*, 108269. [CrossRef]
10. Jin, J.-W.; Jeon, B.-W.; Choi, C.-W.; Kang, K.-W. Multi-Scale Probabilistic Analysis for the Mechanical Properties of Plain Weave Carbon/Epoxy Composites Using the Homogenization Technique. *Appl. Sci.* **2020**, *10*, 6542. [CrossRef]
11. Rodríguez-Ramos, R.; Sabina, F.J.; Guinovart-Díaz, R.; Bravo-Castillero, J. Closed-form expressions for the effective coefficients of a fiber-reinforced composite with transversely isotropic constituents–I. Elastic and square symmetry. *Mech. Mater.* **2001**, *33*, 223–235. [CrossRef]
12. Guinovart-Díaz, R.; Bravo-Castillero, J.; Rodríguez-Ramos, R.; Sabina, F.J.; Rodríguez-Ramos, R.; Bravo-Castillero, J.; Guinovart-Díaz, R. Closed-form expressions for the effective coefficients of a fibre-reinforced composite with transversely isotropic constituents. II: Piezoelectric and hexagonal symmetry. *J. Mech. Phys. Solids* **2001**, *49*, 1463–1479. [CrossRef]
13. Jiang, C.P.; Xu, Y.L.; Cheung, Y.K.; Lo, S.H. A rigorous analytical method for doubly periodic cylindrical inclusions under longitudinal shear and its application. *Mech. Mater.* **2004**, *36*, 225–237. [CrossRef]
14. Eshelby, J.D. The determination of the elastic field of an ellipsoidal inclusion, and related problems. *Proc. R. Soc. Lond. Ser. A* **1957**, *241*, 376–396.
15. Lu, J.-K. *Boundary Value Problems for Analytic Functions*; Series in Pure Mathematics; World Scientific: London, UK, 1994; Volume 16.
16. Artioli, E.; Bisegna, P.; Maceri, F. Effective longitudinal shear moduli of periodic fibre-reinforced composites with radially-graded fibres. *Int. J. Solids Struct.* **2010**, *47*, 383–397. [CrossRef]
17. Shu, W.; Stanciulescu, I. Multiscale homogenization method for the prediction of elastic properties of fiber-reinforced composites. *Int. J. Solids Struct.* **2020**, *203*, 249–263. [CrossRef]
18. Dhimole, V.K.; Chen, Y.; Cho, C. Modeling and Two-Step Homogenization of Aperiodic Heterogenous 3D Four-Directional Braided Composites. *J. Compos. Sci.* **2020**, *4*, 179. [CrossRef]
19. Bisegna, P.; Caselli, F. A simple formula for the effective complex conductivity of periodic fibrous composites with interfacial impedance and applications to biological tissues. *J. Phys. D. Appl. Phys.* **2008**, *41*, 115506. [CrossRef]
20. Godin, Y.A. Effective complex permittivity tensor of a periodic array of cylinders. *J. Math. Phys.* **2013**, *54*, 53505. [CrossRef]
21. Godin, Y.A. Effective properties of periodic tubular structures. *Q. J. Mech. Appl. Math.* **2016**, *69*, 181–193. [CrossRef]
22. Milton, G.W. Bounds on the complex dielectric constant of a composite material. *Appl. Phys. Lett.* **1980**, *37*, 300–302. [CrossRef]

23. Milton, G.W.; Movchan, A.B. A correspondence between plane elasticity and the two-dimensional real and complex dielectric equations in anisotropic media. *Proc. R. Soc. London. Ser. A Math. Phys. Sci.* **1995**, *450*, 293–317.
24. Mackay, T.G.; Lakhtakia, A. Gain and loss enhancement in active and passive particulate composite materials. *Waves Random Complex Media* **2016**, *26*, 553–563. [CrossRef]
25. Guild, M.D.; Garcia-Chocano, V.M.; Kan, W.; Sánchez-Dehesa, J. Enhanced inertia from lossy effective fluids using multi-scale sonic crystals. *AIP Adv.* **2014**, *4*, 124302. [CrossRef]
26. Luong, Q.; Nguyen, M.; TonThat–Long; Tran, D. Complex Shear Modulus Estimation using Integration of LMS/AHI Algorithm. *Int. J. Adv. Comput. Sci. Appl.* **2018**, *9*, 584–589. [CrossRef]
27. Hashin, Z. Thin interphase/imperfect interface in elasticity with application to coated fiber composites. *J. Mech. Phys. Solids* **2002**, *50*, 2509–2537. [CrossRef]
28. Sevostianov, I.; Rodríguez-Ramos, R.; Guinovart-Díaz, R.; Bravo-Castillero, J.; Sabina, F.J. Connections between different models describing imperfect interfaces in periodic fiber-reinforced composites. *Int. J. Solids Struct.* **2012**, *49*, 1518–1525. [CrossRef]
29. Guinovart-Díaz, R.; Rodríguez-Ramos, R.; López-Realpozo, J.C.; Bravo-Castillero, J.; Otero, J.A.; Sabina, F.J.; Lebon, F.; Dumont, S. Analysis of fibrous elastic composites with nonuniform imperfect adhesion. *Acta Mech.* **2016**, *227*, 57–73. [CrossRef]
30. Le, T.-T. Multiscale Analysis of Elastic Properties of Nano-Reinforced Materials Exhibiting Surface Effects. Application for Determination of Effective Shear Modulus. *J. Compos. Sci.* **2020**, *4*, 172. [CrossRef]
31. Bensoussan, A.; Lions, J.; Papanicolaou, G. *Asymptotic Analysis for Periodic Structures*; North-Holland Publishing Company: Amsterdam, The Netherlands, 1978; ISBN 0444851720.
32. Sánchez-Palencia, E. *Non Homogeneous Media and Vibration Theory*; Lecture Notes in Physiscs; Springer: Berlin/Heidelberg, Germany, 1980.
33. Bakhvalov, N.S.; Panasenko, G.P. *Homogenization Averaging Processes in Periodic Media*; Kluwer Academic: Dordrecht, The Netherlands, 1989.
34. Allaire, G.; Briane, M. Multiscale convergence and reiterated homogenisation. *Proc. R. Soc. Edinb. Sect. A Math.* **1996**, *126*, 297–342. [CrossRef]
35. Auriault, J.L.; Boutin, C.; Geindreau, C. *Homogenization of Coupled Phenomena in Heterogenous Media*; Wiley-ISTE: Hoboken, NJ, USA, 2009.
36. Penta, R.; Gerisch, A. Investigation of the potential of asymptotic homogenization for elastic composites via a three-dimensional computational study. *Comput. Vis. Sci.* **2015**, *17*, 185–201. [CrossRef]
37. Ramírez-Torres, A.; Penta, R.; Rodríguez-Ramos, R.; Merodio, J.; Sabina, F.J.; Bravo-Castillero, J.; Guinovart-Díaz, R.; Preziosi, L.; Grillo, A. Three scales asymptotic homogenization and its application to layered hierarchical hard tissues. *Int. J. Solids Struct.* **2018**, *130–131*, 190–198. [CrossRef]
38. Ramírez-Torres, A.; Penta, R.; Rodríguez-Ramos, R.; Grillo, A.; Preziosi, L.; Merodio, J.; Guinovart-Díaz, R.; Bravo-Castillero, J. Homogenized out-of-plane shear response of three-scale fiber-reinforced composites. *Comput. Vis. Sci.* **2019**, *20*, 85–93. [CrossRef]
39. Dong, H.; Zheng, X.; Cui, J.; Nie, Y.; Yang, Z.; Yang, Z. High-order three-scale computational method for dynamic thermo-mechanical problems of composite structures with multiple spatial scales. *Int. J. Solids Struct.* **2019**, *169*, 95–121. [CrossRef]
40. Guinovart-Díaz, R.; López-Realpozo, J.C.; Rodríguez-Ramos, R.; Bravo-Castillero, J.; Ramírez, M.; Camacho-Montes, H.; Sabina, F.J. Influence of parallelogram cells in the axial behaviour of fibrous composite. *Int. J. Eng. Sci.* **2011**, *49*, 75–84. [CrossRef]
41. Kalamkarov, A.L.; Andrianov, I.V.; Pacheco, P.M.C.L.; Savi, M.A.; Starushenko, G.A. Asymptotic Analysis of Fiber-Reinforced Composites of Hexagonal Structure. *J. Multiscale Model.* **2016**, *7*, 1650006–1650270. [CrossRef]
42. Rodríguez-Ramos, R.; Yan, P.; López-Realpozo, J.C.; Guinovart-Díaz, R.; Bravo-Castillero, J.; Sabina, F.J.; Jiang, C.P. Two analytical models for the study of periodic fibrous elastic composite with different unit cells. *Compos. Struct.* **2011**, *93*, 709–714. [CrossRef]
43. Cioranescu, D.; Donato, P. *An Introduction to Homogenization*; Oxford University Press: Oxford, UK, 2000.
44. Muskhelishvili, N.I. *Some Basic Problems of the Mathematical Theory of Elasticity*; Noordhoff: Groningen, Holland, 1953.
45. Grigolyuk, E.I.; Fil'shtinskii, L.A. *Perforated plates and shells*; Nauka: Moscow, Russia, 1970; Volume 1, pp. 6–31.
46. Markushevich, A.I. *Theory of Functions of a Complex Variable*; Silverman, R.A., Ed.; Prentice-Hall: Upper Saddle River, NJ, USA, 1967.
47. Kantorovich, L.V.; Krylov, V.I. *Approximate Methods of Higher Analysis*, 3rd ed.; Interscience: INC, Groningen, The Netherlands, 1964.
48. Van Loocke, M.; Lyons, C.G.; Simms, C.K. Viscoelastic properties of passive skeletal muscle in compression: Stress-relaxation behaviour and constitutive modelling. *J. Biomech.* **2008**, *41*, 1555–1566. [CrossRef]
49. Sharafi, B.; Blemker, S.S. A micromechanical model of skeletal muscle to explore the effects of fiber and fascicle geometry. *J. Biomech.* **2010**, *43*, 3207–3213. [CrossRef]
50. Frolova, K.P.; Vilchevskaya, E.N.; Bauer, S.M.; Müller, W.H. Determination of the shear viscosity of the sclera. *ZAMM J. Appl. Math. Mech. Z. Angew. Math. Mech.* **2019**, *99*, e201800156. [CrossRef]
51. Marcucci, L.; Bondì, M.; Randazzo, G.; Reggiani, C.; Natali, A.N.; Pavan, P.G. Fibre and extracellular matrix contributions to passive forces in human skeletal muscles: An experimental based constitutive law for numerical modelling of the passive element in the classical Hill-type three element model. *PLoS ONE* **2019**, *14*, e0224232. [CrossRef] [PubMed]
52. Kolpakov, A.; Kolpakov, A. Capacity and Transport. In *Contrast Composite Structures: Asymptotic Analysis and Applications*, 1st ed.; CRC Press: Boca Raton, FL, USA, 2009.
53. López-Realpozo, J.C.; Rodríguez-Ramos, R.; Guinovart-Díaz, R.; Bravo-Castillero, J.; Otero, J.A.; Sabina, F.J.; Lebon, F.; Dumont, S.; Sevostianov, I. Effective elastic shear stiffness of a periodic fibrous composite with non-uniform imperfect contact between the matrix and the fibers. *Int. J. Solids Struct.* **2014**, *51*, 1253–1262. [CrossRef]

 *technologies*

Article

Tykhonov Well-Posedness and Convergence Results for Contact Problems with Unilateral Constraints

Mircea Sofonea [1,*,†] and Meir Shillor [2,†]

[1] Laboratoire de Mathématiques et Physique, University of Perpignan, Via Domitia, 52 Avenue Paul Alduy, 66860 Perpignan, France

[2] Department of Mathematics and Statistics, Oakland University, Rochester, MI 48309, USA; shillor@oakland.edu

* Correspondence: sofonea@univ-perp.fr

† The authors contributed equally to this work.

Abstract: This work presents a unified approach to the analysis of contact problems with various interface laws that model the processes involved in contact between a deformable body and a rigid or reactive foundation. These laws are then used in the formulation of a general static frictional contact problem with unilateral constraints for elastic materials, which is governed by three parameters. A weak formulation of the problem is derived, which is in the form of an elliptic variational inequality, and the Tykhonov well-posedness of the problem is established, under appropriate assumptions on the data and parameters, with respect to a special Tykhonov triple. The proof is based on arguments on coercivity, compactness, and lower-semicontinuity. This abstract result leads to different convergence results, which establish the continuous dependence of the weak solution on the data and the parameters. Moreover, these results elucidate the links among the weak solutions of the different models. Finally, the corresponding mechanical interpretations of the conditions and the results are provided. The novelty in this work is the application of the Tykhonov well-posedness concept, which allows a unified and elegant framework for this class of static contact problems.

Keywords: contact problem; unilateral constraint; variational inequality; Tykhonov triple; Tykhonov well-posedness; approximating sequence

Citation: Sofonea, M.; Shillor, M. Tykhonov Well-Posedness and Convergence Results for Contact Problems with Unilateral Constraints. *Technologies* **2021**, *9*, 1. https://dx.doi.org/10.3390/technologies9010001

Received: 19 November 2020
Accepted: 18 December 2020
Published: 24 December 2020

Publisher's Note: MDPI stays neutral with regard to jurisdictional claims in published maps and institutional affiliations.

1. Introduction

Processes of contact between a deformable solid and a foundation are ubiquitous, and they can be found in many industrial settings, in transportation, in various scientific experimental settings, and in everyday life. This is the reason for the very large amount of engineering literature dedicated to the modeling, numerical approximations, and computer simulations of such processes. In addition, indeed, one can find shelf upon shelf of books and journal publications dealing with the myriad aspects of contact processes.

On the other hand, although the Mathematical Theory of Contact Mechanics (MTCM) has expanded substantially in recent years and is quickly maturing because of the substantial mathematical complexity of most models for contact processes, the theory necessarily became more and more abstract. In a way, the gulf between the highly sophisticated abstract theory and the engineering applications became ever more wider. However, the theory yielded also many different effective computer algorithms for the computer approximations of the solutions of the models with various levels of convergence assertions. Thus, the very abstract theory yielded very useful and practical tools for the simulations of contact models.

Mathematically, contact processes are modeled with complex highly nonlinear and often non-smooth boundary value problems, which explains the various mathematical challenges they pose. In particular, their analysis is carried out by using the so-called weak or variational formulation, which is usually in the form of a variational or hemivariational inequality, or more complex differential set-inclusions. The MTCM has provided many existence, uniqueness, and convergence results, as well as the measurability of the solutions when randomness in the system parameters and inputs is allowed. These were obtained by using the mathematical properties of the convexity, monotonicity, lower semicontinuity of various functions and operators, and various fixed point theorems. A sample of MTCM references are, e.g., the books [1–9]. The computation aspects and the related numerical analysis of various models of contact, including numerical simulations, can be find in [10–13], see also the recent survey [14], among a host of many other publications.

A special type of contact problems, which is very challenging mathematically, but is somewhat popular in engineering literature, deals with computational aspects of the models for the processes involved in contact between an elastic solid body and a rigid foundation or surface, the so-called rigid obstacle. This is an idealization of the real process, since there are no perfectly rigid obstacles; however, it is found to be a useful approximation in many applications. Moreover, it leads to a very simple linear complementarity formulation. Indeed, since the obstacle is assumed to be perfectly rigid, the contact conditions are expressed in terms of inequalities for the normal component of the displacement and the stress fields, thus taking into account the non-penetrability of the obstacle or foundation by the body. However, whereas the "classical" formulations is simple, it leads to severe mathematical difficulties, and it took a long time for the MTCM to encompass problems with such a condition. The complementarity condition for the normal surface displacement causes the variational or weak formulation of such problems to be in the form of inequality problems with unilateral constraints. These models may describe a variety of contact settings which arise in the following situations. There is a gap between the surface of the body and the rigid obstacle; there is a thin layer of deformable material that covers the rigid obstacle. Furthermore, the properties of such a thin layer can be elastic, rigid-plastic, or rigid-elastic, for instance. The resulting variational inequalities involve a number of parameters and it is of considerable interest to study the convergence of the solutions with respect to these parameters. Indeed, this allows us to predict the changes in the solutions caused by the perturbations of the data. Moreover, such convergence results establish links between the different models, and justify some of the assumptions made in the modeling of the different physical settings.

The mathematical literature dedicated to general convergence results, within the context of models using differential equations or inclusions, in various settings, function spaces, and under different assumptions is extensive. Such results may be obtained by using different methods and functional arguments, including monotonicity, pseudomonotonicity, compactness, and convexity, among many others. Nevertheless, most of the convergence results in the literature are stated in the following abstract functional framework: Given a functional space X and a problem $\mathcal{P}$ which has a unique solution $u \in X$, a family of approximating problems $\{\mathcal{P}_\theta\}$ is constructed such that, when $u_\theta \in X$ is a solution of Problem $\mathcal{P}_\theta$, then u_θ converges to u in X, as θ converges. A careful analysis of this description reveals that, in practice, we need to complete the functional framework above by describing the following three ingredients: (a) the set I to which the parameter θ belongs; (b) the problem $\mathcal{P}_\theta$ or its sets of solutions, denoted by $\Theta(\theta)$, for each $\theta \in I$; (c) the meaning we give to the convergence of the parameter θ. Collecting these three ingredients, we arrive in a natural way to the concept of *Tykhonov triple*, denoted by $\mathcal{T} = (I, \Theta, \mathcal{C})$, where $\mathcal{C}$ is a set of sequences which governs the convergence of θ.

Basic properties of Tykhonov triples can be found in [15]. There, Tykhonov triples have been used to introduce the general concept of *Tykhonov well-posedness* in metric spaces and then various applications in functional analysis have been described. The Tykhonov well-posedness concept can be applied to the study of a large class of problems: minimization problems, operator equations, fixed point problems, differential equations, inclusions, sweeping processes, and various classes of inequalities as well. It was introduced in the context of optimization problems in the pioneering work [16] and was based on two main ingredients: the existence and uniqueness of the solution to a problem and the convergence of every approximating sequence to this solution. For this reason, it provides a framework in which various convergence results may be stated and proved in a unified way. Tykhonov well-posedness results in the study of viscoplastic constitutive laws, anti-plane shear problems with elastic materials and quasistatic contact problems with elasto-viscoplastic materials can be found in the papers [17–19], respectively.

In this paper, we use Tykhonov triples as the main ingredient of a unified theory of various convergence results in the study of contact problems with unilateral constraints. Our aim in this work is two fold—first, to describe a few mathematical models for the process of contact of a linearly elastic body with unilateral constraints and to prove their unique weak solvability; second, to obtain convergence results with respect to some of the system parameters and to deduce the relationship among the weak solutions of these models. To this end, we prove a Tykhonov well-posedness result, Theorem 1, which is used to establish the two previous tasks. It is seen that this framework and the theorem allow us to obtain these results in a simple, unified, and elegant functional framework.

Following this introduction, the rest of the paper is structured as follows. In Section 2, we describe the interface laws of contact we consider in this manuscript. In Section 3, we present a general mathematical model of static contact, state the assumption on the data, and derive its variational formulation. The latter is in the form of an elliptic quasivariational inequality for the displacement field. Then, in Section 4, we state and prove the Tykhonov well-posedness theorem. We use this result in Sections 5 and 6 in order to obtain various convergence results together with the corresponding mechanical interpretations. Indeed, these convergence results provide a deeper insight into the connections and relations among the various contact models. As an example for the theory, we provide in Section 7 a one-dimensional somewhat simple case of the static contact of a rod with a layered obstacle that, nevertheless, presents the main ideas of our approach without the mathematical complications in two or three dimensions. This, in turn, may be used as a benchmark case for testing numerical methods. Finally, concluding remarks and some future work are provided in Section 8.

2. Interface Laws with Unilateral Constraints

This section presents various interface laws describing the contact process of a deformable body and an obstacle, the so-called foundation. These fall naturally into the conditions in the normal direction and those in the tangential directions. To describe them, we let d belong to the set $\{1, 2, 3\}$ and Ω be a d-dimensional connected domain representing the solid body, and let Γ_D, Γ_N and Γ_C be three relatively open mutually disjoint surfaces such that $\partial\Omega = \overline{\Gamma}_D \cup \overline{\Gamma}_N \cup \overline{\Gamma}_C$. Here, Γ_C denotes the potential contact surface and we let ν be the unit outward normal to Ω. The equalities and inequalities we write below in this section are valid on Γ_C. Nevertheless, for the sake of simplicity, we do not mention it explicitly. We denote by u the displacement field and by σ the stress field in the body. Moreover, we use a dot for the inner product of vectors and the subscripts ν and τ denote the normal component and the tangential part of vectors and tensors, respectively. For instance, the normal and tangential displacements are given by $u_\nu = u \cdot \nu, \quad u_\tau = u - u_\nu \nu$, while the normal and tangential components of the stress field are $\sigma_\nu = (\sigma\nu) \cdot \nu, \quad \sigma_\tau = \sigma\nu - \sigma_\nu \nu$, respectively. We note that the component σ_τ represents the tangential shear or the friction force.

We start with the interface laws in the normal direction, the contact conditions, and consider two different physical settings. In the first one, the foundation is a rigid body and in the second one it is made of a rigid body covered by a layer of deformable material, which may be another material or just the surface asperities.

Contact conditions with a rigid body. First, we assume that the foundation is perfectly rigid, and there is no gap between the deformable body and the foundation, as shown in Figure 1a. Although there are no perfect rigid bodies, the conditions below turn out to be useful in many applied settings. A popular contact condition used both in engineering literature and mathematical publications is the Signorini contact condition, formulated as follows:

$$u_\nu \leq 0, \quad \sigma_\nu \leq 0, \quad \sigma_\nu u_\nu = 0 \quad \text{on } \Gamma_C. \tag{1}$$

This condition was first introduced in [20] and then used in many papers, see e.g., Ref. [7] and the references therein. This condition doesn't allow interpenetration. When $u_\nu < 0$, there is separation between the body and the foundation and (1) implies that $\sigma_\nu = 0$, i.e., the normal stress vanishes. When $u_\nu = 0$, there is contact. Therefore, (1) implies that $\sigma_\nu \leq 0$, i.e., the reaction of the foundation is towards the body. A graphic representation of the the Signorini condition (1) is provided in Figure 2a.

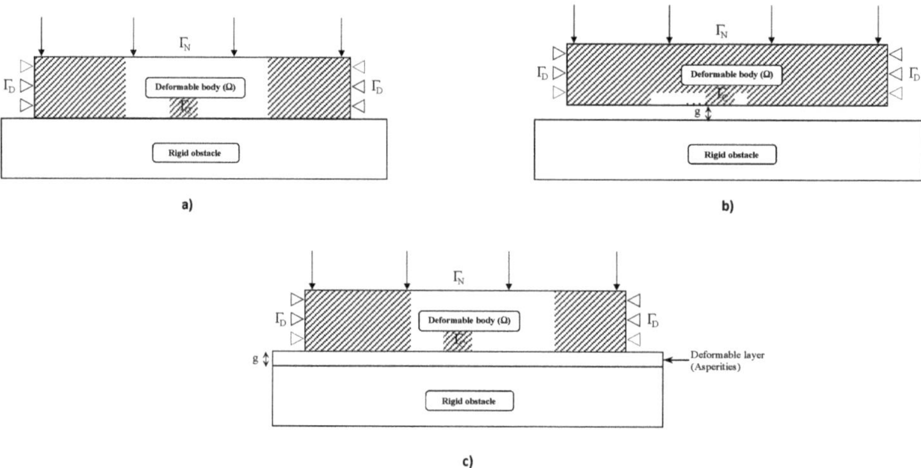

Figure 1. Physical setting: (**a**) contact with a rigid obstacle without gap; (**b**) contact with a rigid obstacle with gap; (**c**) contact with a rigid obstacle covered by a deformable layer.

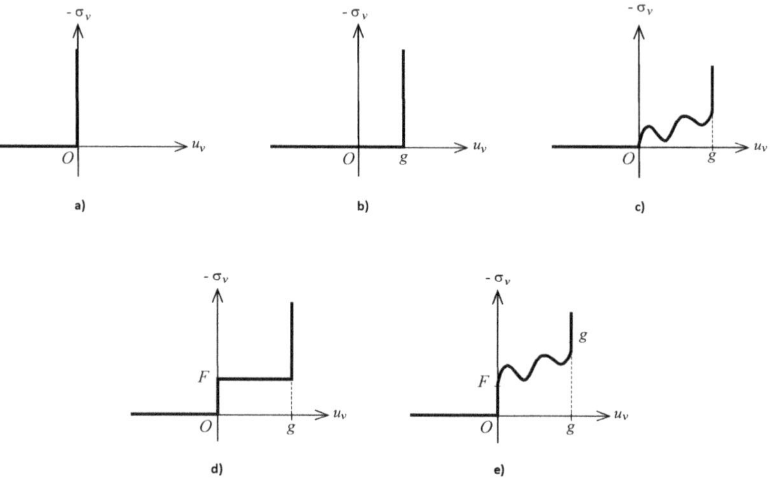

Figure 2. Five contact conditions with unilateral constraints: (**a**) condition (1); (**b**) condition (2); (**c**) condition (7); (**d**) condition (9); (**e**) condition (11).

In the next case, we assume in addition that there is a gap $g > 0$, in the reference configuration, between the body and the foundation, see Figure 1b. Then, the Signorini condition reads

$$u_\nu \leq g, \quad \sigma_\nu \leq 0, \quad \sigma_\nu(u_\nu - g) = 0 \qquad \text{on } \Gamma_C. \tag{2}$$

The mechanical interpretations of (2) is very similar to the case $g = 0$, and, graphically, it is depicted in Figure 2b.

We next consider more complex conditions.

Contact conditions with a rigid body covered by a deformable layer. Consider now the case when the foundation is made of a rigid body covered with a layer of deformable material of thickness $g > 0$. This layer may be just the asperities or a softer material, as is shown is shown in Figure 1c. Then, since the rigid obstacle is impenetrable, we have

$$u_\nu \leq g. \tag{3}$$

Moreover, using the principle of superposition, it follows that the normal stress has an additive decomposition of the form

$$\sigma_\nu = \sigma_\nu^D + \sigma_\nu^R, \tag{4}$$

in which σ_ν^D describes the reaction of the deformable layer and σ_ν^R describes the reaction of the rigid body.

Assume now that the deformable layer has an elastic behavior. Then, for the part σ_ν^D of the normal stress, we use the so-called normal compliance contact condition, which assigns a reactive normal pressure that depends on the interpenetration of the asperities on the body's surface and those of the foundation. Therefore,

$$-\sigma_\nu^D = p(u_\nu) \tag{5}$$

where p is a nonnegative regular function that vanishes for a negative argument. Indeed, when $u_\nu < 0$, there is no contact and the normal pressure vanishes. When $0 \leq u_\nu \leq g$, there is contact and u_ν represents a measure of the interpenetration into the elastic layer. Then, condition (5) shows that the layer exerts on the body a pressure that depends on the penetration. In addition, when $u_\nu = g$, this layer is completely squeezed, and the normal pressure it exerts is $p(g)$. The normal compliance contact condition was first introduced in [21] and since then used in many publications, see e.g., Refs. [13,22–24] and the references therein. On the other hand, for the rigid part of the obstacle, we use the Signorini contact condition with a gap (2). Therefore,

$$\sigma_\nu^R \leq 0, \quad \sigma_\nu^R(u_\nu - g) = 0 \tag{6}$$

and recall that $g > 0$ represents the thickness of the deformable layer. We now gather conditions (3)–(6) and, in this way, we obtain the contact condition

$$u_\nu \leq g, \quad \left.\begin{array}{ll} \sigma_\nu = 0 & \text{if } u_\nu < 0 \\ -\sigma_\nu = p(u_\nu) & \text{if } 0 \leq u_\nu < g \\ -\sigma_\nu \geq p(g) & \text{if } u_\nu = g \end{array}\right\} \quad \text{on } \Gamma_C. \tag{7}$$

A graphic depiction of the contact condition (7) is provided in Figure 2c.

Next, we also consider the case when the deformable layer has a rigid-plastic behavior. In this case, in addition to (3), (4), and (6), we assume that

$$-F \leq \sigma_\nu^D \leq 0, \quad \sigma_\nu^D = \left\{\begin{array}{ll} 0 & \text{if } u_\nu < 0, \\ -F & \text{if } u_\nu > 0. \end{array}\right. \tag{8}$$

Here, F is a given positive traction threshold that may depend on the spatial variable x. Using (8), we have

$$-F < \sigma_v^P \leq 0 \quad \Longrightarrow \quad u_v \leq 0,$$
$$\sigma_v^P = -F \quad \Longrightarrow \quad u_v \geq 0.$$

This shows that the layer does not allow penetration and, therefore, behaves as a rigid body, as far as the inequality $-F < \sigma_v^P \leq 0$ holds. It allows penetration only when the threshold is reached, $\sigma_v^P = -F$ and, then, it offers no additional resistance, as surface plastic flow commences. Thus, conditions (8) model the situation when the deformable layer has a rigid-plastic behavior. Moreover, the function F could be interpreted as the yield limit. Gathering conditions (3), (4), (6), and (8) yields the contact condition

$$u_v \leq g, \qquad \left.\begin{array}{ll} \sigma_v = 0 & \text{if } u_v < 0 \\ -F \leq \sigma_v \leq 0 & \text{if } u_v = 0 \\ \sigma_v = -F & \text{if } 0 < u_v < g \\ \sigma_v \leq -F & \text{if } u_v = g \end{array}\right\} \quad \text{on } \Gamma_C. \tag{9}$$

We may summarize this condition as follows:

(a) If $u_v < 0$, there is no contact and then (8) implies that $\sigma_v^D = 0$, (5) implies that $\sigma_v^R = 0$ and, therefore, equality (4) shows that $\sigma_v = 0$. Thus, the contact traction vanishes, as expected.

(b) If $u_v = 0$, contact has just been established (or is about to be lost) and then (8) implies that $-F \leq \sigma_v^D \leq 0$, (5) implies that $\sigma_v^R = 0$ and, therefore, equality (4) shows that $-F \leq \sigma_v \leq 0$. Thus, the layer behaves as a rigid surface.

(c) If $0 < u_v < g$, there is thus interpenetration into the layer, and then (8) implies that $\sigma_v^D = -F$, and (5) implies that $\sigma_v^R = 0$ and, therefore, equality (4) shows that $\sigma_v = -F$. The layer is in the plastic flow regime.

(d) If $u_v = g$, the layer is completely squashed, and then (8) implies that $\sigma_v^P = -F$, and (5) implies that $\sigma_v^R \leq 0$ and, therefore, equality (4) shows that $\sigma_v \leq -F$.

The contact condition (9) is depicted in Figure 2d. It was used in a number of papers, see, e.g., Ref. [9] and the references therein.

Finally, we consider the case when the deformable layer has a rigid-elastic behavior. In this case, in addition to (3), (4), and (6), we assume that

$$\left.\begin{array}{ll} \sigma_v^D = 0 & \text{if } u_v < 0 \\ -F \leq \sigma_v \leq 0 & \text{if } u_v = 0 \\ -\sigma_v^D = F + p(u_v) & \text{if } u_v > 0 \end{array}\right\}. \tag{10}$$

Condition (10) represents a combination of conditions (5) and (8) in which F is a positive function and p is the normal compliance function; it is positive when the argument is positive and vanishes for a negative argument. Arguments similar to those above show that now the behavior of the deformable layer is rigid-elastic. Here, F could be interpreted as the yield limit of the layer, while the normal compliance function p describes its elastic properties. We now gather (3), (4), (6), and (10) to obtain the following contact condition:

$$u_v \leq g, \qquad \left.\begin{array}{ll} \sigma_v = 0 & \text{if } u_v < 0 \\ -F \leq \sigma_v \leq 0 & \text{if } u_v = 0 \\ -\sigma_v = F + p(u_v) & \text{if } 0 < u_v < g \\ -\sigma_v \geq F + p(u_v) & \text{if } u_v = g \end{array}\right\} \quad \text{on } \Gamma_C. \tag{11}$$

This condition is depicted in Figure 2e.

Comments on the contact conditions (1), (2), (7), (9), and (11). First, these conditions are expressed in terms of unilateral constraints and are governed by the data $g, p,$ and F. Moreover, all of them are described by multivalued relations between the normal displacement and the compressive normal stress, see Figure 2. In addition, there exists a hierarchy among these contact conditions as follows:

(a) Condition (9) can be obtained from condition (11) when the normal compliance function p vanishes, i.e., $p \equiv 0$.

(b) Condition (7) is obtained from condition (11) when the yield limit F vanishes, i.e., $F = 0$.

(c) Condition (2) can be recovered from condition (9), when $p \equiv 0$, from condition (7) when $F = 0$ and from condition (11) when $p \equiv 0$ and $F = 0$.

(d) The Signorini contact condition (1) is obtained from conditions (2), (7), (9), and (11) when $g = 0$.

We conclude that, among the above conditions, condition (11) is the most general one. For this reason, it will play a special role in the next two sections.

Coulomb's law of dry friction. We end this section with the conditions in the tangential directions, also called frictional conditions or friction laws. The simplest one is the so-called frictionless condition in which the tangential part of the stress vanishes. This is an idealization of the process, since even completely lubricated surfaces generate shear resistance to tangential motion. For this reason, we assume in what follows that the tangential traction σ_τ does not vanish on the contact surface, i.e., the contact is with friction.

Frictional contact between solid surfaces without lubrication is usually modeled with a number of variants of the Coulomb law of dry friction. The classical static version of this law, commonly used in frictional contact problems describing the equilibrium states of elastic bodies, is formulated as follows:

$$\|\sigma_\tau\| \le \mu\,|\sigma_\nu|, \quad \sigma_\tau = -\mu\,|\sigma_\nu|\,\frac{u_\tau}{\|u_\tau\|} \quad \text{if} \quad u_\tau \ne 0 \quad \text{on} \quad \Gamma_C. \tag{12}$$

Here, $\mu > 0$ is the coefficient of friction and $\|\sigma_\tau\|$ represents the norm of the friction force. The friction law (12) was intensively studied in the literature; see, for instance, the references in [7]. It shows that, during the contact process, the magnitude of the friction force is bounded by the positive function $\mu\,|\sigma_\nu|$, the friction bound. This is the maximal strength that friction resistance can provide, and above it the surfaces undergo a relative motion. It indicates that the points on the contact surface where the inequality $\|\sigma_\tau\| < \mu\,|\sigma_\nu|$ holds are in the *stick state* since there $u_\tau = 0$. The points of the contact surface where $u_\tau \ne 0$ are in the *slip state*. There, the friction force σ_τ is opposite to the slip u_τ and, moreover, its magnitude equals the magnitude of the friction bound since, in this case, (12) implies that $\|\sigma_\tau\| = \mu\,|\sigma_\nu|$.

We note here that "friction force" is not a force in the usual sense, since friction is only resistance to motion and cannot initiate motion, unlike a "real" force. Although we use the term friction force, "frictional resistance force" is the more accurate term in physics, since it just opposes motion.

We now combine Coulomb's law (12) with each one of the contact conditions (1), (2), (7), (9), or (11), and obtain a specific boundary condition. We note that, when there is separation between the surfaces (i.e., when $u_\nu < 0$ in the case of conditions (1), (7), (9), or (11), and $u_\nu < g$ in the case of condition (2)), then $\sigma_\nu = 0$ and, therefore, the friction bound in (12) vanishes. This, in turn, implies that $\sigma_\tau = 0$, i.e., the friction resistance force vanishes too. This property is realistic from a physical point of view and expresses the compatibility between the contact conditions with unilateral constraints considered above and the Coulomb law of dry friction.

In mathematical publications, and for mathematical reasons mentioned shortly, the classical Coulomb's law of dry friction (12) needs to be modified, and is very often used in its regularized version

$$\|\sigma_\tau\| \le \mu\,|\mathcal{R}\sigma_\nu|, \quad \sigma_\tau = -\mu\,|\mathcal{R}\sigma_\nu|\,\frac{u_\tau}{\|u_\tau\|} \quad \text{if} \quad u_\tau \ne 0 \quad \text{on} \quad \Gamma_C. \tag{13}$$

Here, $\mathcal{R}$ is a continuous regularizing operator that may be considered as the average of the normal stress over a small patch around the contact point. The inclusion of this operator can be traced to [25,26]. As explained in [25], there seems to be some physical justification in considering the normal stress in the friction condition (13) as averaged over a small surface area which contains many asperities, since the physical contact point usually contains many asperities, and the contact surface is rarely smooth. However, the main motivation for such a choice is mathematical, to avoid otherwise insurmountable difficulties. Indeed, in the weak formulation, the regularity of the stress σ does not allow a meaningful definition of the absolute value of the normal stress σ_ν on the boundary. To overcome this difficulty, the operator $\mathcal{R}$ has been introduced in [26]. As an example of such an operator, one may use the convolution of σ with an infinitely differentiable function that has support in a small area that includes the point where the condition is applied.

The constitutive law of an elastic material is such that σ depends explicitly on u, and we may write it as $\sigma = \sigma(u)$, which, in turn, implies that $\sigma_\nu = \sigma_\nu(u)$. Therefore, denoting by R the regularizing operator defined by

$$Ru = \mathcal{R}\sigma_\nu(u),$$

in the case of elastic materials, we can write the regularized friction law (13) as follows:

$$\|\sigma_\tau\| \le \mu\,|Ru|, \quad \sigma_\tau = -\mu\,|Ru|\,\frac{u_\tau}{\|u_\tau\|} \quad \text{if} \quad u_\tau \ne 0 \quad \text{on} \quad \Gamma_C. \tag{14}$$

Details on the regularized friction law (14) can be found in [7] and, therefore, we skip them here. We just mention that in this paper we deal with contact problems for linearly elastic materials and, therefore, we use the regularized version (14) of Coulomb's law of dry friction. The properties of the regularizing operator R will be described in the next section.

3. Main Problem and Variational Formulation

This section presents the physical setting of the contact problem we are interested in, lists the assumption on the problem data, and derives its variational formulation.

Assume that a deformable solid body occupies, in the reference configuration, an open, bounded, and connected set $\Omega \subset \mathbb{R}^d$ ($d = 2, 3$). The boundary $\Gamma = \partial\Omega$ is composed of three relatively closed sets $\overline{\Gamma}_D$, $\overline{\Gamma}_N$ and $\overline{\Gamma}_C$, such that the relatively open sets Γ_D, Γ_N, and Γ_C are mutually disjoint and, moreover, the measure of Γ_D is positive. The body is clamped on Γ_D. Tractions of surface density f_N act on Γ_N and, moreover, body forces of density (per unit volume) f_0 act in Ω. The body can come into contact on Γ_C with another solid, which is called an "obstacle" or "foundation", as shown in Figure 1. Our interest is in the static mechanical equilibrium; the body is assumed to be linearly elastic; and the main interest is in what happens on the contacting surface.

We use bold face letters for vectors and tensors; the outward unit normal on Γ is denoted by ν; the spatial variable is denoted by x and, in order to simplify the notation, we do not indicate explicitly the dependence of the various functions on x.

We denote by $\mathbb{S}^d$ the space of second order symmetric tensors on $\mathbb{R}^d$ and $u : \Omega \to \mathbb{R}^d$ and $\sigma : \Omega \to \mathbb{S}^d$ represent the displacement and the stress fields, respectively. The mathematical model that describes the equilibrium of the elastic body, under the previous mechanical assumptions, consists of the following equations:

$$\sigma = \mathcal{E}\varepsilon(u) \quad \text{in} \quad \Omega, \tag{15}$$

$$\text{Div}\,\sigma + f_0 = 0 \quad \text{in} \quad \Omega. \tag{16}$$

The elastic constitutive law is given in (15) in which $\mathcal{E}$ is the elasticity tensor, and $\varepsilon(u)$ denotes the linearized strain field. The equilibrium Equation (16) describes the static process that is assumed here. Next, the displacement–traction boundary conditions associated with this physical settings are

$$u = 0 \qquad \text{on} \quad \Gamma_D, \tag{17}$$

$$\sigma \nu = f_N \qquad \text{on} \quad \Gamma_N. \tag{18}$$

To complete the model, we add the friction law (14) and one of the contact conditions introduced in Section 2. We recall that condition (11) is the most general one and, therefore, we start by using this contact condition. Since it is governed by the data g, p, F, we denote in what follows by $\mathcal{P}_{gpF}$ the resulting mathematical model. To conclude, the main problem we consider can be stated as follows.

Problem 1. $\mathcal{P}_{gpF}$. *Find a displacement field* $u = u(g, p, F) : \Omega \to \mathbb{R}^d$ *and a stress field* $\sigma = \sigma(g, p, F) : \Omega \to \mathbb{S}^d$ *that satisfy* (15)–(18), (11) *and* (14).

In the variational analysis of this problem, we denote by "$\cdot$", $\|\cdot\|$ and $\mathbf{0}$ the inner product, the Euclidean norm, and the zero element of the spaces $\mathbb{R}^d$ and $\mathbb{S}^d$, respectively. We use the standard notation for the Sobolev and Lebesgue spaces associated with $\Omega \subset \mathbb{R}^d$ and Γ and, for an element $v \in H^1(\Omega)^d$, we usually write v for the trace $\gamma v \in L^2(\Gamma)^d$ of v on Γ. Moreover, we denote by v_ν and v_τ the normal and tangential components of v on the boundary, given by $v_\nu = v \cdot \nu$ and $v_\tau = v - v_\nu \nu$, respectively. We also use the spaces

$$V = \{\, v = (v_i) \in H^1(\Omega)^d : v = \mathbf{0} \text{ on } \Gamma_D \,\},$$
$$Q = \{\, \sigma = (\sigma_{ij}) : \sigma_{ij} = \sigma_{ji} \in L^2(\Omega), \ i,j = 1,\dots,d \,\},$$

which are real Hilbert spaces endowed with the canonical inner products

$$(u, v)_V = \int_\Omega \varepsilon(u) \cdot \varepsilon(v)\, dx, \qquad (\sigma, \tau)_Q = \int_\Omega \sigma \cdot \tau\, dx. \tag{19}$$

Recall that, in (19) and (16), ε and Div represent the deformation and the divergence operators, respectively, i.e.,

$$\varepsilon(u) = (\varepsilon_{ij}(u)), \quad \varepsilon_{ij}(u) = \frac{1}{2}\,(u_{i,j} + u_{j,i}), \quad \operatorname{Div}\sigma = (\sigma_{ij,j}).$$

Here and below, an index that follows a comma denotes the partial derivative with respect to the corresponding component of x, i.e., $u_{i,j} = \partial u_i / \partial x_j$, and the summation convention over a repeated index is used. The associated norms on these spaces are denoted by $\|\cdot\|_V$ and $\|\cdot\|_Q$, respectively. We use $"\to"$ and $"\rightharpoonup"$ to denote the strong and the weak convergence on V and $\mathbf{0}_V$ for the zero element in V. Moreover, it follows from the Sobolev trace arguments that there exists a constant $c_0 > 0$ such that

$$\|v\|_{L^2(\Gamma_C)^d} \le c_0 \|v\|_V, \qquad \forall v \in V. \tag{20}$$

Finally, we recall that, for a regular stress function σ, the following Green's formula holds:

$$\int_\Omega \sigma \cdot \varepsilon(v)\, dx + \int_\Omega \operatorname{Div}\sigma \cdot v\, dx = \int_\Gamma \sigma \nu \cdot v\, dS, \qquad \forall v \in H^1(\Omega)^d. \tag{21}$$

We now list the assumption on the data of the contact problem $\mathcal{P}_{gpF}$. The elasticity tensor $\mathcal{E}$ is symmetric and positively definite, i.e., it satisfies the conditions

$$\begin{cases} \text{(a) } \mathcal{E} = (\mathcal{E}_{ijkl}) : \Omega \times \mathbb{S}^d \to \mathbb{S}^d, \\[4pt] \text{(b) } \mathcal{E}_{ijkl} = \mathcal{E}_{klij} = \mathcal{E}_{jikl} \in L^\infty(\Omega),\ 1 \le i,j,k,l \le d, \\[4pt] \text{(c) There exists } m_\mathcal{E} > 0 \text{ such that} \\ \quad \mathcal{E}\tau \cdot \tau \ge m_\mathcal{E} \|\tau\|^2 \ \forall\, \tau \in \mathbb{S}^d,\ \text{a.e. in } \Omega. \end{cases} \tag{22}$$

The regularization operator R is Lipschitz continuous, i.e.,

$$\begin{cases} R : V \to L^2(\Gamma_C) \text{ and there exists } L_R > 0 \text{ such that} \\ \|Ru - Rv\|_{L^2(\Gamma_C)} \leq L_R \|u - v\|_V \quad \text{for all } u, v \in V. \end{cases} \tag{23}$$

We also assume that the densities of the body forces and surface tractions and the thickness of the deformable layer are such that

$$f_0 \in L^2(\Omega)^d. \tag{24}$$

$$f_N \in L^2(\Gamma_N)^d. \tag{25}$$

$$g \geq 0. \tag{26}$$

Moreover, the normal compliance function p, the yield limit F, and the coefficient of friction μ satisfy the following conditions:

$$\begin{cases} p \colon \Gamma_C \times \mathbb{R} \to \mathbb{R}_+ \text{ and} \\[4pt] \text{(a) there exists } L_p > 0 \text{ such that} \\ \quad |p(x, r_1) - p(x, r_2)| \leq L_p |r_1 - r_2| \\ \quad \text{for all } r_1, r_2 \in \mathbb{R}, \text{ a.e. } x \in \Gamma_C, \\[4pt] \text{(b) } p(\cdot, r) \text{ is measurable on } \Gamma_C \text{ for all } r \in \mathbb{R}, \\[4pt] \text{(c) } p(x, r) = 0 \text{ if and only if } r \leq 0, \text{ a.e. } x \in \Gamma_C. \end{cases} \tag{27}$$

$$F \in L^2(\Gamma_C), \quad F(x) \geq 0 \quad \text{a.e. } x \in \Gamma_C. \tag{28}$$

$$\mu \in L^\infty(\Gamma_C), \quad \mu(x) \geq 0 \quad \text{a.e. } x \in \Gamma_C. \tag{29}$$

Finally, we assume that the following smallness condition holds:

$$c_0^2 L_p + c_0 L_R \|\mu\|_{L^\infty(\Gamma_C)} < m_{\mathcal{E}}, \tag{30}$$

where c_0, L_R, L_p, and $m_{\mathcal{E}}$ are the positive constants in (20), (23), (27), and (22), respectively.

We turn to construct a variational inequality formulation of the problem. To that end, we consider the set $K_g \subset V$, the form $a : V \times V \to \mathbb{R}$, the function $j_{pF} : V \times V \to \mathbb{R}$ and the element $f \in V$ defined by

$$K_g = \{ v \in V : v_\nu \leq g \text{ a.e. on } \Gamma_C \}, \tag{31}$$

$$a(u, v) = \int_\Omega \mathcal{E}\varepsilon(u) \cdot \varepsilon(v) \, dx, \qquad \forall u, v \in V, \tag{32}$$

$$j_{pF}(u, v) = \int_{\Gamma_C} p(u_\nu) v_\nu \, dS + \int_{\Gamma_C} F v_\nu^+ \, dS + \int_{\Gamma_C} \mu |Ru| \, \|v_\tau\| \, dS, \qquad \forall u, v \in V,$$

$$(f, v)_V = \int_\Omega f_0 \cdot v \, dx + \int_{\Gamma_N} f_N \cdot v \, dS, \qquad \forall v \in V, \tag{33}$$

where, here and below, r^+ represents the positive part of r, which is $r^+ = \max\{r, 0\}$.

Next, standard arguments based on the Green formula (21) show that, if (u, σ) is a smooth solution of Problem $\mathcal{P}_{gpF}$ and $v \in K_g$, then

$$\int_{\Omega} \mathcal{E}\varepsilon(u) \cdot (\varepsilon(v) - \varepsilon(u))\, dx = \int_{\Omega} f_0 \cdot (v - u)\, dx + \int_{\Gamma_N} f_N \cdot (v - u)\, dS \tag{34}$$

$$+ \int_{\Gamma_C} \sigma_\nu(v_\nu - u_\nu)\, dS + \int_{\Gamma_C} \sigma_\tau \cdot (v_\tau - u_\tau)\, dS.$$

To deal with the third term on the right-hand side, we rewrite it as

$$\sigma_\nu(v_\nu - u_\nu) = (\sigma_\nu + F + p(u_\nu))(v_\nu - g) + (\sigma_\nu + F + p(u_\nu))(g - u_\nu)$$
$$+ F(u_\nu - v_\nu) + p(u_\nu)(u_\nu - v_\nu),$$

and, using the boundary conditions (11), we deduce that

$$\int_{\Gamma_C} \sigma_\nu(v_\nu - u_\nu)\, dS \geq \int_{\Gamma_C} F(u_\nu^+ - v_\nu^+)\, dS + \int_{\Gamma_C} p(u_\nu)(u_\nu - v_\nu)\, dS. \tag{35}$$

Moreover, using the friction law (14), we find that

$$\int_{\Gamma_C} \sigma_\tau \cdot (v_\tau - u_\tau)\, dS \geq \int_{\Gamma_C} \mu\, |Ru|(\|u_\tau\| - \|v_\tau\|)\, dS. \tag{36}$$

We now combine (34) with inequalities (35) and (36) and obtain

$$\int_{\Omega} \mathcal{E}\varepsilon(u) \cdot (\varepsilon(v) - \varepsilon(u))\, dx + \int_{\Gamma_C} F(v_\nu^+ - u_\nu^+)\, dS + \int_{\Gamma_C} p(u_\nu)(v_\nu - u_\nu)\, dS \tag{37}$$

$$+ \int_{\Gamma_C} \mu\, |Ru|(\|v_\tau\| - \|u_\tau\|)\, dS \geq \int_{\Omega} f_0 \cdot (v - u)\, dx + \int_{\Gamma_N} f_2 \cdot (v - u)\, dS.$$

Finally, we use inequality (37), the notations (32) and (33) and the fact that $u \in K_g$ to obtain the following variational formulation of Problem $\mathcal{P}_{gpF}$, in terms of the displacements.

Problem 2. $\mathcal{P}_{gpF}^V$. *Find a displacement field $u = u(g, p, F)$ such that*

$$u \in K_g, \quad a(u, v - u) + j_{pF}(u, v) - j_{pF}(u, u) \geq (f, v - u)_V, \quad \forall v \in K_g. \tag{38}$$

A function $u = u(g, p, F)$ which satisfies inequality (38) is called *a weak solution* of the contact problem $\mathcal{P}_{gpF}$. Once the existence of a weak solution is found, the stress function can be obtained by using the elastic constitutive law (15).

4. Tykhonov Well-Posedness

In this section, we study the Tykhonov well-posedness of Problem $\mathcal{P}_{gpF}^V$ and, to this end, we start by recalling some of the necessary abstract setting and concepts introduced in [15].

Consider an abstract mathematical object $\mathcal{P}$, called a generic "problem," that is associated with a metric space $(X, \tilde{d})$. Problem $\mathcal{P}$ could be an equation, or a problem of minimization, a fixed point, an inclusion, or an inequality. We associate with Problem $\mathcal{P}$ the concept of "solution", which depends on the context. We also denote by $\mathcal{S}_\mathcal{P} \subset X$ the set of solutions to Problem $\mathcal{P}$. Problem $\mathcal{P}$ has a unique solution iff $\mathcal{S}_\mathcal{P}$ has a unique element, i.e., $\mathcal{S}_\mathcal{P}$ is a singleton. For a nonempty set B, we denote by $\mathcal{S}(B)$ the set of sequences whose elements belong to B, and 2^B is the set of all nonempty subsets of B. The concept of well-posedness for Problem $\mathcal{P}$ is related to the so-called *Tykhonov triple*, defined as follows.

Definition 1. *A Tykhonov triple is a mathematical object of the form $\mathcal{T} = (I, \Theta, \mathcal{C})$, where I is a given nonempty set, $\Theta : I \to 2^X$ and $\mathcal{C}$ is a nonempty subset of the set $\mathcal{S}(I)$.*

Below, we refer to I as the *set of parameters*; the family of sets $\{\Theta(\theta)\}_{\theta \in I}$ represents the family of *approximating sets*; finally, $\mathcal{C}$ defines the *criterion of convergence*.

Definition 2. *Given a Tykhonov triple* $\mathcal{T} = (I, \Theta, \mathcal{C})$, *a sequence* $\{u_n\} \in \mathcal{S}(X)$ *is called an approximating sequence if there exists a sequence* $\{\theta_n\} \in \mathcal{C}$ *such that* $u_n \in \Theta(\theta_n)$ *for each* $n \in \mathbb{N}$.

Definition 3. *Given a Tykhonov triple* $\mathcal{T} = (I, \Theta, \mathcal{C})$, *Problem* $\mathcal{P}$ *is said to be well-posed in the sense of Tykhonov if it has a unique solution, and every approximating sequence converges in* X *to this solution.*

We remark that approximating sequences always exist since, by assumption, $\mathcal{C} \neq \emptyset$ and, moreover, for any sequence $\{\theta_n\} \in \mathcal{C}$ and any $n \in \mathbb{N}$, the set $\Theta(\theta_n)$ is not empty. In addition, the concept of approximating sequence depends on the Tykhonov triple $\mathcal{T}$ and, for this reason, we use the terminology "$\mathcal{T}$-approximating sequence". As a consequence, the concept of well-posedness depends on the Tykhonov triple $\mathcal{T}$ and, therefore, we refer to it as "well-posedness with respect to $\mathcal{T}$" or "$\mathcal{T}$-well-posedness," for short.

We turn now on the well-posedness of Problem $\mathcal{P}_{gpF}^V$ and, to this end, we consider the Tykhonov triple $\mathcal{T} = (I, \Theta, \mathcal{C})$, defined as follows:

$$I = \{\, \theta = (\widetilde{g}, \varepsilon) \in \mathbb{R}^2 \; : \; g \geq 0, \; \varepsilon \geq 0 \,\}, \tag{39}$$

$$\Theta(\theta) = \{\, u \in K_{\widetilde{g}} \; : \; a(u, v - u) + j_{pF}(u, v) - j_{pF}(u, u) \tag{40}$$
$$+ \varepsilon \|v - u\|_V \geq (f, v - u)_V \quad \forall v \in K_{\widetilde{g}} \,\} \quad \text{for } \theta = (\widetilde{g}, \varepsilon) \in I,$$

$$\mathcal{C} = \{\, \{\theta_n\} \subset \mathcal{S}(I) \; : \; \theta_n = (g_n, \varepsilon_n) \; \forall n \in \mathbb{N}, : g_n \to g, \; \varepsilon_n \to 0 \,\}. \tag{41}$$

Here, $\widetilde{g}$ represents a potential thickness and the set $K_{\widetilde{g}}$ is defined by (31), replacing g with $\widetilde{g}$. We next note that, for mathematical reasons, we introduce a positive parameter ε in the definition of the Tykhonov triple (39)–(41). Convenient choices of this parameter allow us to obtain various convergence results to the solution of the variational inequality (38), as we show in Section 5.

Our main result in this section is the following:

Theorem 1. *Assume that* (22)–(30) *holds. Then, Problem* $\mathcal{P}_{gpF}^V$ *is well-posed with respect to the Tykhonov triple* (39)–(41).

The $\mathcal{T}$-well-posedness of Problem $\mathcal{P}_{gpF}^V$ can be established by using the general results on the well-posedness of variational-hemivariational inequalities in [27]. Nevertheless, the statement of the results there requires additional definitions and preliminaries and, therefore, for the convenience of the reader, we present here a direct proof of Theorem 1, which is structured in four steps, as follows.

Proof. (i) *Existence of a unique solution of Problem* $\mathcal{P}_{gpF}^V$. First, we remark that K_g, defined by (31), is a closed, nonempty, and convex set in V. Next, assumptions (22) on the elasticity tensor show that the bilinear form $a : V \times V \to \mathbb{R}$, defined by (32), is symmetric, continuous, and coercive. More precisely, it satisfies the inequality

$$a(v, v) \geq m_{\mathcal{E}} \|v\|_V^2, \quad \forall v \in V. \tag{42}$$

In addition, using the assumptions (23), (27)–(29), it follows that the functional $j_{pF}(u, \cdot) : V \to \mathbb{R}$ is convex and continuous. Then, the inequalities (20) and (23) imply that

$$j_{pF}(u_1, v_2) - j_{pF}(u_1, v_1) + j_{pF}(u_2, v_1) - j_{pF}(u_1, v_2) \tag{43}$$

$$\leq (c_0^2 L_p + c_0 L_R \|\mu\|_{L^\infty(\Gamma_C)}) \|u_1 - u_2\|_V \|v_1 - v_2\|_V, \quad \forall u_1, u_2, v_1, v_2 \in V.$$

Inequalities (42) and (43) combined with the smallness assumption (30) allow us to use Theorem 3.7 in [28] to deduce the unique solvability of Problem $\mathcal{P}^V_{gpF}$.

(ii) *Weak convergence of approximating sequences.* Assume that $\{u_n\}$ is a $\mathcal{T}$-approximating sequence. Then, using Definition 2, we deduce that there exists a sequence $\{\theta_n\} \in \mathcal{C}$ such that $u_n \in \Theta(\theta_n)$ for each $n \in \mathbb{N}$. Therefore, definitions (40), (41), and (31) imply that

$$u_n \in K_{g_n}, \quad a(u_n, v - u_n) + j_{pF}(u_n, v) - j_{pF}(u_n, u_n) \tag{44}$$
$$+\varepsilon_n \|v - u_n\|_V \geq (f, v - u_n)_V \quad \forall v \in K_{g_n},$$

for each $n \in \mathbb{N}$, where

$$K_{g_n} = \{ v \in V : v_v \leq g_n \quad \text{a.e. on } \Gamma_C \}, \tag{45}$$

and, moreover,

$$g_n \to g, \tag{46}$$
$$\varepsilon_n \to 0 \tag{47}$$

as $n \to \infty$.

Let $n \in \mathbb{N}$ be fixed. We choose $v = 0_V$ in (44) and, since $j(u_n, 0_V) = 0$, $j(u_n, u_n) \geq 0$, we find that

$$a(u_n, u_n) \leq \varepsilon_n \|u_n\|_V + (f, u_n)_V.$$

Next, inequality (42) implies that

$$\|u_n\|_V \leq \frac{1}{m_\varepsilon}(\varepsilon_n + \|f\|_V),$$

and, using (47), we obtain that the sequence $\{u_n\}$ is bounded in V. This, in turn, implies that there exists an element $\tilde{u} \in V$ and a subsequence of $\{u_n\}$, still denoted by $\{u_n\}$, such that

$$u_n \rightharpoonup \tilde{u} \quad \text{in } V. \tag{48}$$

It follows that $u_n \to \tilde{u}$ a.e. on Γ_C and, using the definitions (45), (31) combined with the convergence (46), we deduce that

$$\tilde{u} \in K_g. \tag{49}$$

Let now $v \in K_g$ and, for each $n \in \mathbb{N}$, consider the element v_n defined by

$$v_n = \begin{cases} \dfrac{g_n}{g} v & \text{if } g > 0, \\[2mm] v & \text{if } g = 0. \end{cases}$$

Then, it is straightforward to see that $v_n \in K_{g_n}$ and, moreover,

$$v_n \to v \quad \text{in } V. \tag{50}$$

We now use (44) to see that

$$j_{pF}(u_n, v_n) - j_{pF}(u_n, u_n) \geq a(u_n, u_n) - a(u_n, v_n) + (f, v_n - u_n)_V.$$

then we pass to the lower limit in this inequality and use the convergences (48) and (50), the compactness of the trace operator and the properties of the form a and the function j_{pF} to find that

$$j_{pF}(\tilde{u}, v) - j_{pF}(\tilde{u}, \tilde{u}) \geq \liminf a(u_n, u_n) - a(\tilde{u}, v) + (f, v - \tilde{u})_V. \tag{51}$$

Next, since $a(u_n - \tilde{u}, u_n - \tilde{u}) \geq 0$, we obtain

$$a(u_n, u_n) \geq 2a(u_n, \tilde{u}) - a(\tilde{u}, \tilde{u}),$$

which implies that

$$\liminf a(u_n, u_n) \geq a(\tilde{u}, \tilde{u}). \tag{52}$$

Combining (49), (51), and (52) yields

$$\tilde{u} \in K_g, \qquad a(\tilde{u}, v - \tilde{u}) + j_{pF}(\tilde{u}, v) - j_{pF}(\tilde{u}, \tilde{u}) \geq (f, v - \tilde{u})_V,$$

which shows that $\tilde{u}$ is a solution of Problem $\mathcal{P}^V_{gpF}$. We now use the uniqueness of the solution of this problem to deduce that $\tilde{u} = u$. This equality and a standard argument imply that the whole sequence $\{u_n\}$ convergences weakly to u in V, i.e.,

$$u_n \rightharpoonup u \quad \text{in } V. \tag{53}$$

(iii) *Strong convergence of approximating sequences.* For each $n \in \mathbb{N}$, we consider the element $\tilde{u}_n$ defined by

$$\tilde{u}_n = \begin{cases} \dfrac{g_n}{g} u & \text{if } g > 0 \\[2mm] u & \text{if } g = 0. \end{cases}$$

Then, $\tilde{u}_n \in K_{g_n}$ and, moreover,

$$\tilde{u}_n \to u \quad \text{in } V. \tag{54}$$

In addition, it follows from (44) that

$$a(u_n, u_n - \tilde{u}_n) \leq j_{pF}(u_n, \tilde{u}_n) - j_{pF}(u_n, u_n) + \varepsilon_n \|\tilde{u}_n - u_n\|_V - (f, \tilde{u}_n - u_n)_V. \tag{55}$$

We now use the coercivity of the form a (42) to find that

$$m_{\mathcal{E}} \|u_n - \tilde{u}_n\|_V^2 \leq a(u_n - \tilde{u}_n, u_n - \tilde{u}_n) = a(u_n, u_n - \tilde{u}_n) - a(\tilde{u}_n, u_n - \tilde{u}_n),$$

and then (55) yields

$$\begin{aligned} m_{\mathcal{E}} \|u_n - \tilde{u}_n\|_V^2 \leq\ & j_{pF}(u_n, \tilde{u}_n) - j_{pF}(u_n, u_n) + \varepsilon_n \|\tilde{u}_n - u_n\|_V \\ & - (f, \tilde{u}_n - u_n)_V - a(\tilde{u}_n, u_n - \tilde{u}_n). \end{aligned}$$

Next, we pass to the limit in this inequality and use the convergences (54), (53), and (47) to deduce that

$$u_n - \tilde{u}_n \to 0_V \quad \text{in } V. \tag{56}$$

Finally, we combine (54) and (56) and obtain

$$u_n \to u \quad \text{in } V, \tag{57}$$

which concludes the proof of this step.

(iv) *The proof.* It follows from step (i) that Problem $\mathcal{P}^V_{gpF}$ has a unique solution, and it follows from the step (iii) that every $\mathcal{T}$-approximating sequence converges in V to this solution. These two facts combined with Definition 3 show that Problem $\mathcal{P}^V_{gpF}$ is well-posed with respect to the Tykhonov triple (39)–(41), and this concludes the proof. $\square$

5. A Convergence Result for Perturbation of g, p, and F

In this section, we use the well-posedness result provided by Theorem 1 to obtain perturbation convergence results. To that end, we assume in what follows that (22)–(30) hold and denote by $u = u(g, p, F)$ the solution of Problem $\mathcal{P}^V_{gpF}$ in Theorem 1. For each

$n \in \mathbb{N}$, we consider a perturbation g_n, p_n, F_n of the data g, p, F which satisfy conditions (26)–(28), respectively, denoted in what follows by $(26)_n$, $(27)_n$, $(28)_n$. We also denote by L_{p_n} the Lipschitz constant of the function p_n and we suppose that the following smallness assumption holds for every n,

$$c_0^2 L_{p_n} + c_0 L_R \|\mu\|_{L^\infty(\Gamma_C)} < m_{\mathcal{E}}. \tag{58}$$

We denote by $j_{p_n F_n} : V \times V \to \mathbb{R}$ the function

$$j_{p_n F_n}(u, v) = \int_{\Gamma_C} p_n(u_v) v_v \, dS + \int_{\Gamma_C} F_n v_v^+ \, dS + \int_{\Gamma_C} \mu \, |Ru| \, \|v_\tau\| \, dS, \quad \forall u, v \in V \tag{59}$$

and, using notation (45), we consider the following variational problem.

Problem 3. $\mathcal{P}^V_{g_n p_n F_n}$. *Find a displacement field* $u_n = u(g_n, p_n, F_n)$ *such that*

$$u_n \in K_{g_n}, \quad a(u_n, v - u_n) + j_{p_n F_n}(u_n, v) - j_{p_n F_n}(u_n, u_n) \geq (f, v - u_n)_V, \quad \forall v \in K_{g_n}. \tag{60}$$

Note that Problem $\mathcal{P}^V_{g_n p_n F_n}$ is obtained from Problem $\mathcal{P}^V_{gpF}$ by replacing the data g, p, F with the perturbed data g_n, p_n, F_n, respectively. Moreover, it follows from Theorem 1 that, under the assumption stated above, Problem $\mathcal{P}^V_{g_n p_n F_n}$ has a unique solution $u_n = u(g_n, p_n, F_n)$, for each $n \in \mathbb{N}$. To study the convergence of this solution as $n \to \infty$, we consider in what follows the following additional assumptions:

$$g_n \to g \quad \text{as } n \to \infty. \tag{61}$$

$$F_n \to F \quad \text{in } L^2(\Gamma_C) \quad \text{as } n \to \infty. \tag{62}$$

$$\left\{ \begin{array}{l} \text{(a) For each } n \in \mathbb{N} \text{ there exists } \omega_n \geq 0 \text{ such that} \\ \qquad |p_n(x, r) - p(x, r)| \leq \omega_n \quad \text{for all } r \in \mathbb{R}, \text{ a.e. } x \in \Gamma_C. \\ \text{(b) } \omega_n \to 0 \quad \text{as } n \to \infty. \end{array} \right. \tag{63}$$

The main result in this section is the following.

Theorem 2. *Assume that* (22)–(30), $(26)_n$, $(27)_n$, $(28)_n$, (58), (61)–(63) *hold. Then, the solutions* u_n *of Problems* $\mathcal{P}^V_{g_n p_n F_n}$ *converge to the solution* u *of Problem* $\mathcal{P}^V_{gpF}$*, which is*

$$u_n = u(g_n, p_n, F_n) \to u = u(g, p, F) \quad \text{in } V \quad \text{as } n \to \infty. \tag{64}$$

Proof. Let $n \in \mathbb{N}$ and $v \in V$. Then, using the definitions (59) and (33), we find that

$$j_{p_n F_n}(u_n, v) - j_{p_n F_n}(u_n, u_n) - j_{pF}(u_n, v) + j_{pF}(u_n, u_n)$$
$$= \int_{\Gamma_C} \left(p_n(u_{nv}) - p(u_{nv}) \right)(v_v - u_{nv}) \, dS + \int_{\Gamma_C} (F_n - F)(v_v^+ - u_{nv}^+) \, dS$$
$$\leq \int_{\Gamma_C} |p_n(u_{nv}) - p(u_{nv})| |v_v - u_{nv}| \, dS + \int_{\Gamma_C} |F_n - F| |v_v^+ - u_{nv}^+| \, dS$$
$$\leq \int_{\Gamma_C} |p_n(u_{nv}) - p(u_{nv})| \|v - u_n\| \, dS + \int_{\Gamma_C} |F_n - F| \|v - u_n\| \, dS,$$

and then assumption (63) (a) implies that

$$j_{p_n F_n}(u_n, v) - j_{p_n F_n}(u_n, u_n) - j_{pF}(u_n, v) + j_{pF}(u_n, u_n)$$
$$\leq \omega_n \|v - u_n\|_{L^1(\Gamma_C)^d} + \|F_n - F\|_{L^2(\Gamma_C)} \|v - u_n\|_{L^2(\Gamma_C)^d}.$$

This inequality combined with the continuity of the embedding $L^2(\Gamma_C)^d \subset L^1(\Gamma_C)^d$, and the trace inequality (20) shows that there exists a constant α, which does not depend on n, such that

$$j_{p_n F_n}(u_n, v) - j_{p_n F_n}(u_n, u_n) - j_{pF}(u_n, v) + j_{pF}(u_n, u_n) \tag{65}$$
$$\leq \alpha\big(\omega_n + \|F_n - F\|_{L^2(\Gamma_C)}\big)\|v - u_n\|_V.$$

Next, we use the notation

$$\varepsilon_n = \alpha\big(\omega_n + \|F_n - F\|_{L^2(\Gamma_C)}\big), \tag{66}$$

$$\theta_n = (g_n, \varepsilon_n). \tag{67}$$

Then, using (65) and (66), we find that

$$j_{p_n F_n}(u_n, v) - j_{p_n F_n}(u_n, u_n) \leq j_{pF}(u_n, v) - j_{pF}(u_n, u_n) + \varepsilon_n \|v - u_n\|_V.$$

and, therefore, (60) implies that

$$u_n \in K_{g_n}, \quad a(u_n, v - u_n) + j_{pF}(u_n, v) - j_{pF}(u_n, u_n) \tag{68}$$
$$+\varepsilon_n\|v - u_n\|_V \geq (f, v - u_n)_V, \quad \forall v \in K_{g_n}.$$

We now combine (67), (68), and (40) to see that $u_n \in \Theta(\theta_n)$ and, since assumptions (61), (62), and (63) (b) imply that $g_n \to g$ and $\varepsilon_n \to 0$, it follows from (41) that $\{\theta_n\} \subset \mathcal{C}$. We conclude from Definition 2 that $\{u_n\}$ is a $\mathcal{T}$-approximating sequence for Problem $\mathcal{P}_{gpF}^V$. The convergence (64) is now a direct consequence of Theorem 1 and Definition 3. $\quad\square$

Note that convergence (64) expresses the continuous dependence of the solution of Problem $\mathcal{P}_{gpF}^V$ with respect to the data g, p, and F. Besides the mathematical interest in this result, it is important from the mechanical and applications points of view since it shows that small perturbation in the thickness g, the yield limit F, and the normal compliance function p imply small changes in the weak solution of the contact problem $\mathcal{P}_{gpF}$.

6. Additional Convergence Results

We turn now to some special cases of the general convergence result (64), related to the different boundary conditions mentioned in Section 2, for which we present additional mechanical interpretations. To this end, we consider the following contact problems:

Problem 4. $\mathcal{P}_{gF}$. *Find a displacement field* $u = u(g, F) : \Omega \to \mathbb{R}^d$ *and a stress field* $\sigma = \sigma(g, F) : \Omega \to \mathbb{S}^d$ *which satisfy* (15)–(18), (9), *and* (14).

Problem 5. $\mathcal{P}_{gp}$. *Find a displacement field* $u = u(g, p) : \Omega \to \mathbb{R}^d$ *and a stress field* $\sigma = \sigma(g, p) : \Omega \to \mathbb{S}^d$ *which satisfy* (15)–(18), (7), *and* (14).

Problem 6. $\mathcal{P}_g$. *Find a displacement field* $u = u(g) : \Omega \to \mathbb{R}^d$ *and a stress field* $\sigma = \sigma(g) : \Omega \to \mathbb{S}^d$, *which satisfy* (15)–(18), (2), *and* (14).

Problem 7. $\mathcal{P}$. *Find a displacement field* $u = u : \Omega \to \mathbb{R}^d$ *and a stress field* $\sigma = \sigma : \Omega \to \mathbb{S}^d$ *which satisfy* (15)–(18), (1) *and* (14).

Using the relationship between the contact conditions (1), (2), (7), (9), and (11) discussed in Section 2, we have:

(a)   Problem $\mathcal{P}_{gF}$ is a particular case of Problem $\mathcal{P}_{gpF}$, obtained when $p \equiv 0$.
(b)   Problem $\mathcal{P}_{gp}$ is a particular case of Problem $\mathcal{P}_{gpF}$, obtained when $F = 0$.

(c)   Problem $\mathcal{P}_g$ is a particular case of Problem $\mathcal{P}_{gp}$, obtained when $p \equiv 0$, a particular case of Problem $\mathcal{P}_{gF}$ obtained when $F = 0$, and a particular case of Problem $\mathcal{P}_{gpF}$, obtained when $p \equiv 0$ and $F = 0$.

(d)   Problem $\mathcal{P}$ is a particular case of Problems $\mathcal{P}_{gpF}$, $\mathcal{P}_{gp}$ and $\mathcal{P}_{gF}$ obtained when $g = 0$, for any p and F.

Therefore, using the notation

$$K = \{\, v \in V \,:\, v_\nu \leq 0 \text{ a.e. on } \Gamma_C \,\}, \tag{69}$$

$$j_p(u, v) = \int_{\Gamma_C} p(u_\nu)v_\nu \, dS + \int_{\Gamma_C} \mu \, |R\sigma_\nu(u)| \, \|v_\tau\| \, dS, \qquad \forall\, u,\, v \in V, \tag{70}$$

$$j_F(u, v) = \int_{\Gamma_C} F v_\nu^+ \, dS + \int_{\Gamma_C} \mu \, |R\sigma_\nu(u)| \, \|v_\tau\| \, dS, \qquad \forall\, u,\, v \in V, \tag{71}$$

$$j(u, v) = \int_{\Gamma_C} \mu \, |R\sigma_\nu(u)| \, \|v_\tau\| \, dS, \qquad \forall\, u,\, v \in V, \tag{72}$$

the variational formulations of these problems represent particular cases of the Problem $\mathcal{P}_{gpF}^V$ and are as follows:

Problem 8. $\mathcal{P}_{gF}^V$. *Find a displacement field* $u = u(g, F)$ *such that*

$$u \in K_g, \quad a(u, v - u) + j_F(u, v) - j_F(u, u) \geq (f, v - u)_V, \quad \forall\, v \in K_g.$$

Problem 9. $\mathcal{P}_{gp}^V$. *Find a displacement field* $u = u(g, p)$ *such that*

$$u \in K_g, \quad a(u, v - u) + j_p(u, v) - j_p(u, u) \geq (f, v - u)_V, \quad \forall\, v \in K_g.$$

Problem 10. $\mathcal{P}_g^V$. *Find a displacement field* $u = u(g)$ *such that*

$$u \in K_g, \quad a(u, v - u) + j(u, v) - j(u, u) \geq (f, v - u)_V, \quad \forall\, v \in K_g.$$

Problem 11. $\mathcal{P}^V$. *Find a displacement field* u *such that*

$$u \in K, \quad a(u, v - u) + j(u, v) - j(u, u) \geq (f, v - u)_V, \quad \forall\, v \in K.$$

We now make the somewhat weaker assumption

$$c_0 L_R \|\mu\|_{L^\infty(\Gamma_C)} < m_\mathcal{E}, \tag{73}$$

and note that, if (58) holds with $L_{p_n} > 0$, then (73) holds too. Then, the unique solvability of the variational problems above is provided in the following result.

Corollary 1. *Assume that* (22)–(25), (29) *hold. Then:*

(a)   *Under assumptions* (26), (28), *and* (73) *Problem* $\mathcal{P}_{gF}^V$ *has a unique solution* $u = u(g, F)$.

(b)   *Under assumptions* (26), (27), *and* (30) *Problem* $\mathcal{P}_{gp}^V$ *has a unique solution* $u = u(g, p)$.

(c)   *Under assumptions* (26) *and* (73), *Problem* $\mathcal{P}_g^V$ *has a unique solution* $u = u(g)$.

(d)   *Under assumption* (73), *Problem* $\mathcal{P}^V$ *has a unique solution* u.

Corollary 1 is a direct consequence of the unique solvability of the variational problem $\mathcal{P}_{gpF}$, guaranteed by Theorem 1 and Definition 3. Moreover, under assumptions (23), (27), (28), and (29), it is straightforward to check that

$$j_{pF}(u, v) = j_p(u, v) = j_F(u, v) = j(u, v), \qquad \forall\, u,\, v \in K.$$

Therefore, with the notation in Corollary 1, we have

$$u(0, p, F) = u(0, F) = u(0, p) = u(0) = u. \tag{74}$$

To proceed with the analysis, we introduce the following assumptions:

$$g_n \to 0 \quad \text{as } n \to \infty. \tag{75}$$

$$F_n \to 0 \quad \text{in } L^2(\Gamma_C) \quad \text{as } n \to \infty. \tag{76}$$

$$\begin{cases} \text{(a) For each } n \in \mathbb{N} \text{ there exists } \omega_n \geq 0 \text{ such that} \\ \quad |p_n(x, r)| \leq \omega_n \quad \text{for all } r \in \mathbb{R}, \text{ a.e. } x \in \Gamma_C. \\ \text{(b) } \omega_n \to 0 \quad \text{as } n \to \infty. \end{cases} \tag{77}$$

Then, as a direct consequence of Theorem 2 and equalities (74), we obtain the following convergence results.

Corollary 2. *Assume that* (22)–(25), (29) *hold. Then:*

(a) *Under assumptions* (26), (28), $(26)_n$, $(27)_n$, $(28)_n$, (58), (61), (62), (77), *the solutions* $u(g_n, p_n, F_n)$ *of Problems* $\mathcal{P}^V_{g_n p_n F_n}$ *converge to the solution* $u(g, F)$ *of Problem* $\mathcal{P}^V_{gF}$, *which is*

$$u(g_n, p_n, F_n) \to u(g, F) \quad \text{in } V \quad \text{as } n \to \infty. \tag{78}$$

(b) *Under assumptions* (26), (27), $(28)_n$, (30), $(26)_n$, $(27)_n$, (58), (61), (63), *and* (76), *the solution* $u(g_n, p_n, F_n)$ *of Problem* $\mathcal{P}^V_{g_n p_n F_n}$ *converges to the solution* $u(g, p)$ *of Problem* $\mathcal{P}^V_{gp}$, *which is*

$$u(g_n, p_n, F_n) \to u(g, p) \quad \text{in } V \quad \text{as } n \to \infty. \tag{79}$$

(c) *Under assumptions* (26), $(26)_n$, $(27)_n$, $(28)_n$, (58), (61), (76), *and* (77), *the solution* $u(g_n, p_n, F_n)$ *of Problem* $\mathcal{P}^V_{g_n p_n F_n}$ *converges to the solution* $u(g)$ *of Problem* $\mathcal{P}^V_{g}$, *which is*

$$u(g_n, p_n, F_n) \to u(g) \quad \text{in } V \quad \text{as } n \to \infty. \tag{80}$$

(d) *Under assumptions* $(26)_n$, (27), $(27)_n$, (28), $(28)_n$, (58), (62), (63), *and* (75), *the solution* $u(g_n, p_n, F_n)$ *of Problem* $\mathcal{P}^V_{g_n p_n F_n}$ *converges to the solution* u *of Problem* $\mathcal{P}^V$, *which is*

$$u(g_n, p_n, F_n) \to u \quad \text{in } V \quad \text{as } n \to \infty. \tag{81}$$

(e) *Under assumptions* (26), $(26)_n$, (28), $(28)_n$, (61), (62), *and* (73), *the solution* $u(g_n, F_n)$ *of Problem* $\mathcal{P}^V_{g_n F_n}$ *converges to the solution* $u(g, F)$ *of Problem* $\mathcal{P}^V_{gF}$, *which is*

$$u(g_n, F_n) \to u(g, F) \quad \text{in } V \quad \text{as } n \to \infty. \tag{82}$$

(f) *Under assumptions* (26), $(26)_n$, $(28)_n$, (61), (73), *and* (76), *the solution* $u(g_n, F_n)$ *of Problem* $\mathcal{P}^V_{g_n F_n}$ *converges to the solution* $u(g)$ *of Problem* $\mathcal{P}^V_{g}$, *which is*

$$u(g_n, F_n) \to u(g) \quad \text{in } V \quad \text{as } n \to \infty. \tag{83}$$

(g) *Under assumptions* $(26)_n$, (28), $(28)_n$, (62), (73), *and* (75), *the solution* $u(g_n, F_n)$ *of Problem* $\mathcal{P}^V_{g_n F_n}$ *converges to the solution* u *of Problem* $\mathcal{P}^V$, *which is*

$$u(g_n, F_n) \to u \quad \text{in } V \quad \text{as } n \to \infty. \tag{84}$$

(h) *Under assumptions* (26), (27), $(26)_n$, $(27)_n$, (30), (58), (61), *and* (63), *the solution* $u(g_n, p_n)$ *of Problem* $\mathcal{P}^V_{g_n p_n}$ *converges to the solution* $u(g, p)$ *of Problem* $\mathcal{P}^V_{gp}$, *which is*

$$u(g_n, p_n) \to u(g, p) \quad \text{in } V \quad \text{as } n \to \infty. \tag{85}$$

(i)   *Under assumptions (26), $(26)_n$, $(27)_n$, (58), (61), and (77), the solution $u(g_n, p_n)$ of Problem $\mathcal{P}^V_{g_n p_n}$ converges to the solution $u(g)$ of Problem $\mathcal{P}^V_g$, which is*

$$u(g_n, p_n) \to u(g) \quad \text{in } V \quad \text{as } n \to \infty. \tag{86}$$

(j)   *Under assumptions $(26)_n$, (27), $(27)_n$, (58), (63), and (75), the solution $u(g_n, p_n)$ of Problem $\mathcal{P}^V_{g_n p_n}$ converges to the solution u of Problem $\mathcal{P}^V$, which is*

$$u(g_n, p_n) \to u \quad \text{in } V \quad \text{as } n \to \infty. \tag{87}$$

(k)   *Under assumptions (26), $(26)_n$, (61), and (73), the solution $u(g_n)$ of Problem $\mathcal{P}^V_{g_n}$ converges to the solution u of Problem $\mathcal{P}^V_g$, which is*

$$u(g_n) \to u(g) \quad \text{in } V \quad \text{as } n \to \infty. \tag{88}$$

(l)   *Under assumptions $(26)_n$, (73), and (75), the solution $u(g_n)$ of Problem $\mathcal{P}^V_{g_n}$ converges to the solution u of Problem $\mathcal{P}^V$, which is*

$$u(g_n) \to u \quad \text{in } V \quad \text{as } n \to \infty. \tag{89}$$

Each one of the convergences above has an appropriate mechanical interpretation. Moreover, they indicate how such problems with these interface or boundary conditions can be approximated by the related problems.

First, the convergences (82), (85), and (88) establish the continuous dependence of the weak solutions of Problems $\mathcal{P}_{gF}$, $\mathcal{P}_{gp}$, and $\mathcal{P}_g$, respectively, with respect to the data. Note that in this case the convergences hold between solutions of problems constructed with the same interface law, but with different data. In contrast, the rest of the results in Corollary 2 lead to convergence of the weak solutions of contact problems that have a different feature, since they are formulated in terms of different interface laws. Indeed, for instance, we list the following:

(a) In the particular case when $g_n = g$ and $p_n = p$, the convergence (79) becomes

$$u(g, p, F_n) \to u(g, p) \quad \text{in } V \quad \text{as } \quad F_n \to 0 \quad \text{in } L^2(\Gamma_C).$$

This shows that the weak solution of the contact problem with a rigid foundation covered by an elastic layer, Figure 2c, can be approached by the solution of a the contact problem with a foundation made by a rigid body covered by a layer of rigid-elastic material, Figure 2e, when the yield limit F of this layer converges to zero, so the layer becomes fully elastic.

(b) In the particular case when $F_n = F$, the convergence (84) and equalities (74) imply that

$$u(g_n, F) \to u(0, F) = u \quad \text{in } V \quad \text{as } \quad g_n \to 0.$$

This shows that the weak solution of the contact problem with a rigid body, Figure 2a, can be approached by the solution of the contact problem with a foundation made by a rigid body covered by a layer of rigid-plastic material, Figure 2d, when the thickness of this layer converges to zero.

We note that, in addition to the mathematical interest in these convergence results (which asserts the stability of the solution), they are very important from the mechanical point of view, since they allows us to establish the links among the different contact models. Indeed, these results show that, for small values of some of the parameters, we can replace, that is, approximate as closely as we wish, some of the more complex models by simpler ones.

7. A One-Dimensional Example

This section illustrates our theoretical results and studies a representative one-dimensional example, that of a static elastic beam in contact with a two-layered foundation. We chose it since it is easier to explain the main ideas of this work but without the complications that arise in two or three dimensions. Thus, we consider a version of Problem $\mathcal{P}_{gF}$, where the elastic beam of length $l = 1\,[m]$ is rigidly attached at $x = 0$ and may come in contact, under the action of a force density (per unit length) $f\,[kg/s^2]$, with a foundation at $x = 1$. The foundation has a deformable layer of the rigid-plastic type of thickness $g\,[m]$, which is attached to a rigid body underneath. In the notation above, we have $\Omega = (0,1)$, $\Gamma_D = \{x = 0\}$, $\Gamma_N = \varnothing$, $\Gamma_C = \{x = 1\}$. The setting is depicted in Figure 3.

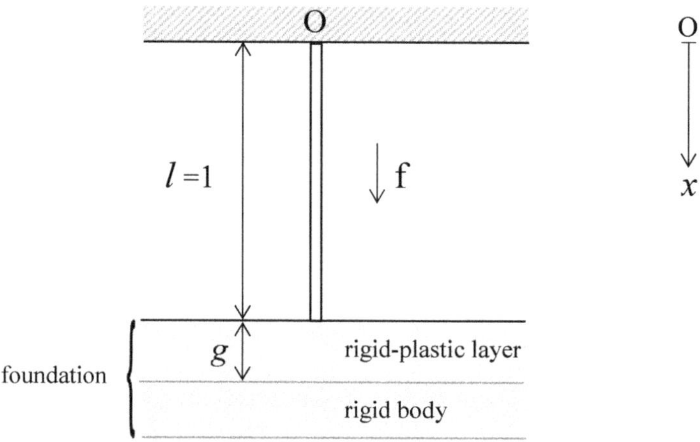

Figure 3. Physical setting.

We denote by $u = u(x)$ [m] the displacement, and then the linearized strain field is given by $\varepsilon(u) = u'$ (dimensionless), where, here and below, the prime denotes the derivative with respect to $x \in [0,1]$. We denote by Y [kg/m s^2] the Young modulus of the rod's material, A [m^2] the cross sectional area of the rod, and then $E = YA$ [kg m/s^2] is the effective (1D) Young modulus. The stress in the rod is given by $\sigma(x)$ [kg m/s^2], and within linearized elasticity, $\sigma = Eu'$. For the sake of simplicity, we assume that $f \in \mathbb{R}$ does not depend on the spatial variable.

The statement of the problem of *static contact between an elastic rod and a rigid-plastic foundation* is the following.

Problem 12. $\mathcal{P}_{gF}^{1d}$. *Find a displacement field $u\colon [0,1] \to \mathbb{R}$ and a stress field $\sigma\colon [0,1] \to \mathbb{R}$, such that*

$$\sigma(x) = E\,u'(x) \quad \text{for } x \in (0,1), \tag{90}$$

$$\sigma'(x) + f = 0 \quad \text{for } x \in (0,1), \tag{91}$$

$$u(0) = 0, \tag{92}$$

$$u(1) \le g, \quad
\begin{cases}
\sigma(1) = 0 & \text{if } u(1) < 0 \\
-F \le \sigma(1) \le 0 & \text{if } u(1) = 0 \\
\sigma(1) = -F & \text{if } 0 < u(1) < g \\
\sigma(1) \le -F & \text{if } u(1) = g
\end{cases} \tag{93}$$

Here, $F\,[kg\,m/sec^2]$ is the rigid-plastic material yield limit, assumed to be positive. One can combine Equations (1) and (2) into $Eu'' + f = 0$; however, we write it in this way to conform to the formulation of the abstract problems above.

We proceed to the the variational formulation and analysis of Problem $\mathcal{P}^{1d}_{gF}$. To that end, we use the space

$$V = \{\, v \in H^1(0,1) \ : \ v(0) = 0 \,\},$$

and the set of admissible displacement fields is defined by

$$K_g = \{\, u \in V \ : u(1) \le g \,\}.$$

Then, the variational form of Problem $\mathcal{P}^{1d}$, obtained using integration by parts, is as follows.

Problem 13. $\mathcal{P}^{1d-V}_{gF}$. *Find a displacement field* $u \in K_g$ *such that*

$$\int_0^1 Eu'(v' - u')\,dx + Fv(1)^+ - Fu(1)^+ \ge \int_0^1 f(v - u)\,dx \quad \forall\, v \in K_g. \tag{94}$$

The existence of a unique solution to Problem $\mathcal{P}^{1d-V}_{gF}$ follows from Corollary 1(a). However, since the example is "simple", direct calculations allow us to solve Problem $\mathcal{P}^{1d}_{gF}$ and obtain closed form solutions. As was noted above, these may be used to calibrate and verify numerical algorithms for realistic engineering problems. It is found that there are four different possible cases that depend on the relationship between f and F. We describe each one and its corresponding mechanical interpretation.

(a) **The case** $f < 0$. The body force acts away from the foundation and then the solution of Problem $\mathcal{P}^{1d}$ is given by

$$\begin{cases} \sigma(x) = f(1 - x), \\[4pt] u(x) = \frac{f}{E}\left(1 - \tfrac{1}{2}x\right)x, \end{cases} \qquad \forall\, x \in [0,1]. \tag{95}$$

In this case, as is to be expected since there is no contact, $u(1) < 0$ and $\sigma(1) = 0$. Since there is separation between the rod's end and the foundation, there is no reaction at $x = 1$. This case corresponds to Figure 4a.

(b) **The case** $0 \le f < 2F$. The force pushes the rod towards the foundation and the solution of Problem $\mathcal{P}^{1d}$ is given by

$$\begin{cases} \sigma(x) = \frac{f}{2}(1 - 2x), \\[4pt] u(x) = \frac{f}{2E}(1 - x)x, \end{cases} \qquad \forall\, x \in [0,1]. \tag{96}$$

We have $u(1) = 0$ and $-F < \sigma(1) \le 0$, which shows that the rod is in contact with the foundation, just touching it, and the reaction of the foundation is towards the rod. Nevertheless, there is no penetration, since the magnitude of the stress at $x = 1$ is under the yield limit F and, therefore, the rigid-plastic layer behaves like a rigid layer. This case is depicted in Figure 4b.

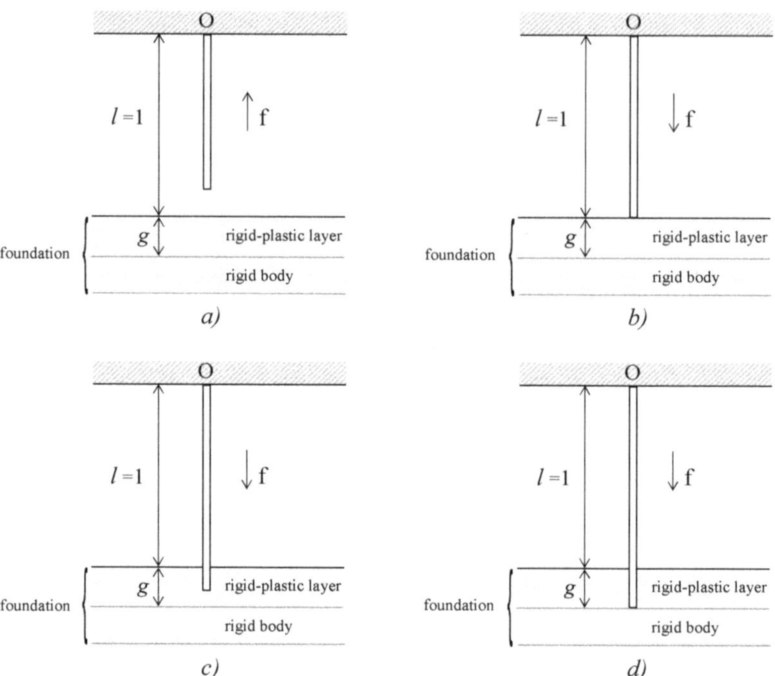

Figure 4. The four cases of contact between the rod and the foundation: (**a**) The case $f < 0$; (**b**) the case $0 \le f < 2F$; (**c**) the case $2F \le f < 2Eg + 2F$; (**d**) the case $2Eg + 2F \le f$.

(c) **The case** $2F \le f < 2Eg + 2F$. In this case, the force is sufficiently large to cause the penetration of the rod's end into the rigid-plastic layer. The solution of Problem $\mathcal{P}^{1d}$ is given by

$$
\begin{cases}
\sigma(x) = f(1 - x) - F, \\
u(x) = \frac{f}{2E}(2 - x)x - \frac{F}{E}x,
\end{cases}
\quad \forall x \in [0, 1]. \tag{97}
$$

We have $0 \le u(1) < g$ and $-\sigma(1) = F$. This, indeed, shows that the stress at $x = 1$ reached the yield limit and, therefore, there is penetration into the rigid-plastic layer which now behaves plastically. Nevertheless, the penetration is partial and $u(1) < g$. This case is shown in Figure 4c.

(d) **The case** $2Eg + 2F \le f$. Here, the applied force is sufficient to make the whole layer plastic. The solution of Problem $\mathcal{P}^{1d}$ is given by

$$
\begin{cases}
\sigma(x) = \frac{f}{2}(1 - 2x) + Eg, \\
u(x) = \frac{f}{2E}(1 - x)x + g,
\end{cases}
\quad \forall x \in [0, 1]. \tag{98}
$$

We have $u(1) = g$ and $\sigma(1) \le -F$, which shows that the rigid-plastic layer is completely penetrated and the displacement of the point $x = 1$ reaches the rigid body. The magnitude of the reaction in this point is larger than the yield limit F since, besides the reaction of the rigid-plastic layer, there is also the reaction of the rigid body, which becomes active in this case. This case is depicted in Figure 4d.

The analytic forms (95)–(98) of the solution in the four cases show clearly the continuous dependence of the solution on the data F and g, which is the content of Corollary

2(e)–(g). For instance, denote in what follows by (u_n, σ_n) the solution to Problem $\mathcal{P}^{1d}_{gF}$ for $g = g_n > 0$ and $F_n = F > 0$, for all $n \in \mathbb{N}$. Then, it follows from (95)–(98) that

$$u_n(x) \to \widetilde{u}(x), \qquad \sigma_n(x) \to \widetilde{\sigma}(x) \quad \text{for all } x \in [0,1] \tag{99}$$

where $\widetilde{u} \colon [0,1] \to \mathbb{R}$ and $\widetilde{\sigma} \colon [0,1] \to \mathbb{R}$ are the functions defined by

$$\begin{cases} \widetilde{\sigma}(x) = \frac{f}{2}(1 - 2x), \\ \widetilde{u}(x) = \frac{f}{2E}(1 - x)x, \end{cases} \qquad \forall\, x \in [0,1]$$

if $f < 0$, and

$$\begin{cases} \widetilde{\sigma}(x) = \frac{f}{2}(1 - 2x), \\ \widetilde{u}(x) = \frac{f}{2E}(1 - x)x, \end{cases} \qquad \forall\, x \in [0,1]$$

if $f \geq 0$. On the other hand, it is easy to see that the couple $(\widetilde{u}, \widetilde{\sigma})$ is the solution to the Signorini problem without a gap, that is:

Problem 14. $\mathcal{P}^{1d}$. *Find a displacement field $u \colon [0,1] \to \mathbb{R}$ and a stress field $\sigma \colon [0,1] \to \mathbb{R}$, which satisfy (90)–(92) and $u(1) \leq 0$, $\sigma(1) \leq 0$, $\sigma(1)u(1) = 0$.*

Therefore, the convergence (99) represents a validation of the convergence result (84).

8. Conclusions

In this paper, we considered a general mathematical model, actually a framework that describes the equilibrium of a system of a linearly elastic body that is in contact with a number of different types of foundations. The model includes four important particular cases, which depend on the assumptions on the system and its parameters. The variational formulation of the general model is in the form of an elliptic quasivariational inequality for the displacement field of the contacting body. We prove the well-posedness of this inequality with respect to a specific Tykhonov triple, and we use this result to deduce convergence results of the solutions with respect to the parameters. Finally, this unified theory for dealing with the variants of the model with the various contact conditions and these convergence results provides the framework that clearly shows the links and relationships among the weak solutions of the different contact settings and conditions. We also provide a "simple" example with four cases that make the theory transparent and the various concepts about the continuous dependence of the solutions on the data easier to follow. This example has interest in and of itself as it may be used as a benchmark for computer simulations of "real" problems.

Our results in this work can be extended in several directions. First, a more general elastic constitutive law of the form $\sigma = \mathcal{F}\varepsilon(u)$, in which $\mathcal{F}$ is a strongly monotone Lipschitz continuous nonlinear operator, can be studied. In such a case, the proof of Theorem 1 can be recovered by using pseudomonotonicity arguments. Second, the dependence of the solution on the density of body forces, density of surface tractions, and coefficient of friction can be obtained, under appropriate assumptions, by using an appropriate choice of ε_n in (44). Extensions to quasistatic contact problems with viscoelastic or viscoplastic materials or to contact problems with nonsmooth interface boundary conditions can also be obtained. Steps in this direction have been made in [19,29], where the concept of Tykhonov triple and Tykhonov well-posedness have been used. It also may be of considerable interest to extend the current methodology to include additional processes on the contacting surfaces such as adhesion of damage [7]. Finally, numerical analysis and computer simulations of these theoretical convergence results would be welcome.

Besides the novelty of the results in this paper, we illustrate the use of the new mathematical tools in the variational analysis of contact problems with unilateral constraints. This is an additional reinforcement of one of the main features of the Mathematical Theory of Contact Mechanics, which is the substantial cross fertilization between the models and applications, on one hand, and the nonlinear functional analysis, on the other hand.

Author Contributions: Conceptualization, M.S. (Mircea Sofonea) and M.S. (Meir Shillor); methodology, M.S. (Mircea Sofonea) and M.S. (Meir Shillor); formal analysis, M.S. (Mircea Sofonea); writing—original draft preparation, M.S. (Mircea Sofonea); writing—review and editing, M.S. (Meir Shillor); supervision, M.S. (Meir Shillor); funding acquisition, M.S. (Mircea Sofonea) and M.S. (Meir Shillor). All authors have read and agreed to the published version of the manuscript.

Funding: This research was funded by European Union's Horizon 2020 Research and Innovation Program under the Marie Sklodowska-Curie Grant Agreement No. 823731 CONMECH.

Conflicts of Interest: The authors declare no conflict of interest. The funders had no role in the design of the study; in the collection, analyses, or interpretation of data; in the writing of the manuscript, or in the decision to publish the results.

References

1. Capatina, A. Advances in Mechanics and Mathematics. In *Variational Inequalities and Frictional Contact Problems*; Springer: New York, NY, USA, 2014; Volume 31.
2. Duvaut G.; Lions, J.-L. *Inequalities in Mechanics and Physics*; Springer: Berlin, Germany, 1976.
3. Eck, E.; Jarušek, J.; Krbec, M. Pure and Applied Mathematics. In *Unilateral Contact Problems: Variational Methods and Existence Theorems*; CRC Press: New York, NY, USA, 2005; Volume 270.
4. Naniewicz Z.; Panagiotopoulos, P.D. *Mathematical Theory of Hemivariational Inequalities and Applications*; Marcel Dekker, Inc.: New York, NY, USA, 1995.
5. Panagiotopoulos, P.D. *Inequality Problems in Mechanics and Applications*; Birkhäuser: Boston, MA, USA, 1985.
6. Panagiotopoulos, P.D. *Hemivariational Inequalities, Applications in Mechanics and Engineering*; Springer: Berlin, Germany, 1993.
7. Shillor, M.; Sofonea, M.; Telega, J.J. *Models and Analysis of Quasistatic Contact*; Lecture Notes in Physics; Springer: Berlin, Germany, 2004; Volume 655.
8. Sofonea, M.; Matei, A. *Mathematical Models in Contact Mechanics*; London Mathematical Society Lecture Note Series 398; Cambridge University Press: Cambridge, UK, 2012.
9. Sofonea, M.; Migórski, S. Pure and Applied Mathematics. In *Variational-Hemivariational Inequalities with Applications*; Chapman & Hall: Boca Raton, MA, USA; CRC Press: London, UK, 2018.
10. Zavarise, G.; Wriggers, P. (Eds.) *Trends in Computational Contact Mechanics*; Lecture Notes in Applied and Computational Mechanics; Springer: Berlin, Germany, 2011; Volume 58.
11. Glowinski, R. *Numerical Methods for Nonlinear Variational Problems*; Springer: New York, NY, USA, 1984.
12. Han, W.; Sofonea, M. Studies in Advanced Mathematics. In *Quasistatic Contact Problems in Viscoelasticity and Viscoplasticity*; American Mathematical Society: Providence, RI, USA; International Press: Somerville, MA, USA, 2002; Volume 30.
13. Kikuchi, N.; Oden, J.T. *Contact Problems in Elasticity: A Study of Variational Inequalities and Finite Element Methods*; SIAM: Philadelphia, PA, USA, 1988.
14. Han, W.; Sofonea, M. Numerical analysis of hemivariational inequalities in Contact Mechanics. *Acta Numer.* **2019**, *28*, 175–286.
15. Sofonea, M.; Xiao, Y.B. Tykhonov triples, Well-posedness and convergence results. *Carphatian J. Math.* **2020**, in press.
16. Tykhonov, A.N. On the stability of functional optimization problems. *USSR Comput. Math. Math. Phys.* **1966**, *6*, 631–634.
17. Sofonea, M.; Tarzia, D.A. On the Tykhonov Well-posedness of an antiplane shear problem. *Mediterr. J. Math.* **2020**, *17*, 1–21, doi:10.1007/s00009-020-01577-5.
18. Sofonea, M. Tykhonov Well-posedness of a rate-type constitutive law. *Mech. Res. Commun.* **2020**, *108*, 103566, doi:10.1016/j.mechrescom. 2020.103566.
19. Sofonea, M.; Xiao, Y.B. Tykhonov Well-posedness of a viscoplastic contact problem. *J. Evol. Equ. Control Theory* **2020**, *9*, 1167–1185.
20. Signorini, A. Sopra alcune questioni di elastostatica. *Soc. Ital. Prog. Sci.* **1933**, *21*, 143–148.
21. Oden J.T.; Martins, J.A.C. Models and computational methods for dynamic friction phenomena. *Comput. Methods Appl. Mech. Eng.* **1985**, *52*, 527–634.
22. Klarbring, A.; Mikelič, A.; Shillor, M. Frictional contact problems with normal compliance. *Int. J. Eng. Sci.* **1988**, *26*, 811–832.
23. Klarbring, A.; Mikelič, A.; Shillor, M. On friction problems with normal compliance. *Nonlinear Anal. Theory Methods Appl.* **1989**, *13*, 935–955.
24. Martins, J.A.C.; Oden, J.T. Existence and uniqueness results for dynamic contact problems with nonlinear normal and friction interface laws. *Nonlinear Anal. Theory Methods Appl.* **1987**, *11*, 407–428.
25. Oden, J.T.; Pires, E. Nonlocal and nonlinear friction laws and variational principles for contact problems in elasticity. *J. Appl. Mech.* **1983**, *50*, 67–82.
26. Duvaut, G. Loi de frottement non locale. *J. Méc. Thé. Appl.* **1982**, 73–78.
27. Sofonea, M.; Xiao, Y.B. Tykhonov Well-posedness of elliptic Variational-Hemivariational Inequalities. *Electron. J. Differ. Equ.* **2019**, *64*, 64.

28. Sofonea, M.; Matei, A. Advances in Mechanics and Mathematics. In *Variational Inequalities with Applications: A Study of Antiplane Frictional Contact Problems*; Springer: New York, NY, USA, 2009; Volume 18.
29. Liu, Z.; Sofonea, M.; Xiao, Y.B. Tykhonov Well-posedness of a frictionless unilateral contact problem. *Math. Mech. Solids* **2020**, *25*, 1294–1311.

 *technologies*

Article

A Model of Damage for Brittle and Ductile Adhesives in Glued Butt Joints

Maria Letizia Raffa [1,*] , Raffaella Rizzoni [2] and Frédéric Lebon [3]

[1] Laboratory QUARTZ EA7393, Supméca, 93400 Saint-Ouen-sur-Seine, France
[2] Department of Engineering, University of Ferrara, 44122 Ferrara, Italy; raffaella.rizzoni@unife.it
[3] Aix-Marseille Université, CNRS, Centrale Marseille, LMA, 13453 Marseille, France; lebon@lma.cnrs-mrs.fr
* Correspondence: maria-letizia.raffa@supmeca.fr

Abstract: The paper presents a new analytical model for thin structural adhesives in glued tube-to-tube butt joints. The aim of this work is to provide an interface condition that allows for a suitable replacement of the adhesive layer in numerical simulations. The proposed model is a nonlinear and rate-dependent imperfect interface law that is able to accurately describe brittle and ductile stress–strain behaviors of adhesive layers under combined tensile–torsion loads. A first comparison with experimental data that were available in the literature provided promising results in terms of the reproducibility of the stress–strain behavior for pure tensile and torsional loads (the relative errors were less than 6%) and in terms of failure strains for combined tensile–torsion loads (the relative errors were less than 14%). Two main novelties are highlighted: (i) Unlike the classic spring-like interface models, this model accounts for both stress and displacement jumps, so it is suitable for soft and hard adhesive layers; (ii) unlike classic cohesive zone models, which are phenomenological, this model explicitly accounts for material and damage properties of the adhesive layer.

Keywords: adhesive layer; butt joint; mode-I; mixed-mode; damage evolution; analytical solution

Citation: Raffa, M.L.; Rizzoni, R.; Lebon, F. A Model of Damage for Brittle and Ductile Adhesives in Glued Butt Joints. *Technologies* **2021**, *9*, 19. https://doi.org/10.3390/technologies9010019

Received: 9 February 2021
Accepted: 3 March 2021
Published: 6 March 2021

Publisher's Note: MDPI stays neutral with regard to jurisdictional claims in published maps and institutional affiliations.

1. Introduction

Within the last decades, adhesive bonding became a very common assembly technique in many industrial sectors, such as aeronautical (e.g., in composite aircraft to bond the stringers to fuselage and wing skins to stiffen the structures against buckling [1]), civil (e.g., in glass-fiber-reinforced polymer pultruded beams [2] or in carbon-fiber-reinforced polymer beams [3]), automotive (e.g., in both closures and structural modules [4]), and biomedical engineering (to fix implants in bone tissue in orthopedic or dentistry surgery [5]), as an alternative to conventional joining techniques, such as welding and riveting [6]. Adhesive bonding provides several advantages, including reduced stress concentrations, higher corrosion resistance, water tightness, and the ability to join materials with dissimilar properties. Moreover, this technique is increasingly chosen by the transport industry (automotive and aeronautics) because it allows the production of lighter structures, thus reducing CO_2 emissions. Nevertheless, adhesive bonding still presents some disadvantages. One of the main concerns limiting the use of adhesive joints is their long-life durability when exposed to service conditions [4]. Corrosion and aging may cause micro-cracking phenomena that can be measured via non-destructive techniques [7]. Another drawback is represented by the multifactorial and multiscale nature of the damage phenomena occurring in the adhesive joints, which make it more complex to predict their strength.

In some structural polymeric adhesives, the tensile stress–strain behavior is typically characterized by an initial linear-elastic phase, followed by softening and rupture. This nonlinear constitutive behavior suggests that a micro-cracking process could occur: pre-existing microcracks, generated by the adhesive preparation (manufacturing, thermal treatment, etc.) and initially present in the linear-elastic phase, propagate during the softening phase, causing debonding and failure [8].

Generally, tube-to-tube butt joints are used to experimentally characterize the mechanical properties of structural adhesives under combined tensile–torsion loads [9–11]. Despite numerous experimental studies on this subject, it is still not possible to univocally define the damage/failure behavior of adhesive layers (see the disadvantages listed above). For this reason, a modeling approach can be very useful, and that is what this work proposes.

In numerical modeling, it is often suitable to avoid a volumetric description of the adhesive layer in order to limit problems that can be involved (e.g., a mesh size that is too small, mesh dependency, too large of a number of degrees of freedom, and too long of a computational time). The classic strategy used for modeling damage in adhesive-bonded joints is based on cohesive zone models (CZMs) [6], which are described by a traction–separation (TS) law across the cohesive surface. Several TS laws of different shapes (i.e., bilateral, trapezoidal, polynomial) have been proposed (see [12,13] and the references therein),and they adequately describe the global response of adhesive-bonded joints [14–17]. However, a crucial drawback of CZMs is that they adopt a phenomenological approach, and thus, the model parameters describing the damage/failure behavior of adhesives are not based on their physical properties (e.g., material properties, geometry).

To overcome this drawback, for the past few years, the authors have been working on alternative TS laws, issued by an imperfect interface approach combining continuum damage mechanics and asymptotic homogenization. These imperfect interface laws have already established their effectiveness in taking into account the micromechanical properties of the adhesive, such as anisotropy [18], micro-cracking, and roughness [19]. Moreover, they can describe the behavior of *hard* adhesive layers (as stiff as adherents) in which both stress and displacement jumps occur [20–22]. Recently, the authors provided a new hard imperfect interface model accounting for micro-cracking damage [23] via an evolution law that is directly related to the mechanical properties of the adhesive.

As a novelty, this paper aims to apply the hard imperfect interface model cited above to the case of adhesive layers in glued butt joints submitted to combined tensile–torsion loads. In detail, a tube-to-tube butt joint configuration is chosen in order to provide an analytical law that, once implemented in a finite element code, can simulate standard characterization tests for structural adhesives.

The presentation of the analytical interface model and its original validation by comparison with experimental data by Murakami et al. [9] are the subjects of this paper, which is organized as follows: the analytical model is presented in Section 2; its numerical implementation together with the chosen experimental data from [9] are detailed in Section 3; the results are illustrated in Section 4, and finally, conclusions and perspectives of future work are highlighted in the summary.

2. Analytical Method for Damage Prediction

After introducing the equilibrium problem of the tube-to-tube butt joint, a classical solution is first introduced, corresponding to the perfect contact between the adherents and modeling a very rigid adhesive. Next, a generalization of the classical solution is proposed, taking into account the presence of a very thin deformable adhesive. The latter is described by a model of an imperfect interface proposed in [22]. Micro-cracking damage within the adhesive is described by using the Kachanov–Sevostianov (KS) model for micro-cracked materials [24,25]. Damage evolution is accounted for by the evolution law obtained in [23] via an asymptotic method.

2.1. Classical Solution for Perfect Contact between the Adherents

The butt-joint specimen is composed of two identical cylindrical adherents that are joined together. The lower basis of the specimen is fixed, and the upper one is subjected to combined tensile force F and torque T, as shown in Figure 1, where the dimensions of the adherents are also shown. Under the simplifying hypotheses of perfect adhesion, small strains, and linear-elastic material behavior, the stress tensor in the adherents is given by the classical relation

$$\sigma(x_1, x_2) = -\frac{T}{I_0}\big(-x_2(\mathbf{e}_1 \odot \mathbf{e}_3) + x_1(\mathbf{e}_2 \odot \mathbf{e}_3)\big) + \frac{F}{A}(\mathbf{e}_3 \otimes \mathbf{e}_3), \tag{1}$$

where $\mathbf{e}_i$ is the versor of the i axis, $i = 1, 2, 3$, the symbol $\odot$ is taken to denote the symmetric dyadic product of vectors, and A and I_0 are the cross-sectional area and the polar moment of inertia, respectively. The stress tensor (1) is divergence-free, and the surface forces $\sigma\mathbf{n}$ vanish on the lateral surface of the cylinder, as $\mathbf{n}$ is the outward normal to the lateral surface. The resultant vertical force and torque on the upper basis of the cylinder balance the applied force F and torque T, respectively, so equilibrium is ensured.

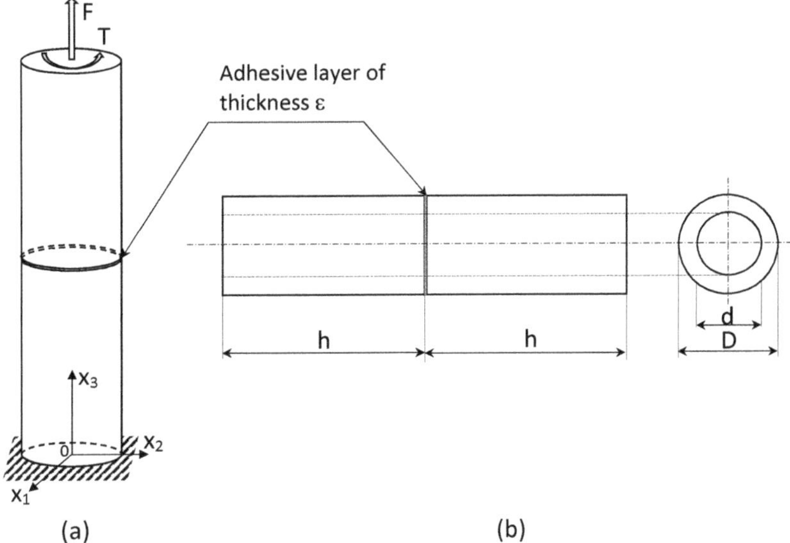

Figure 1. (**a**) Sketch of the tube-to-tube joint with the loading configuration. (**b**) Longitudinal and traversal sections with dimensions.

Assuming the adherents to be made of the same linearly elastic isotropic material with Young's modulus E, Poisson's ratio ν, and shear modulus $G = E/(2(1 + \nu))$, the homogeneous displacement field in the adherents corresponding to the stress (1) is

$$\mathbf{u}^0(x_1, x_2, x_3) = \Big(-\frac{F}{A}\frac{\nu}{E}x_1 - \frac{T}{GI_0}x_2x_3\Big)\mathbf{e}_1 + \Big(-\frac{F}{A}\frac{\nu}{E}x_2 + \frac{T}{GI_0}x_1x_3\Big)\mathbf{e}_2 + \frac{F}{EA}x_3\mathbf{e}_3. \tag{2}$$

2.2. Generalized Equilibrium Solution for Imperfect Contact between the Adherents

The displacement field (2) is appropriate for a specimen made of two identical adherents that are perfectly joined. To take into account the presence of a very thin elastic adhesive without describing it geometrically in a numerical model, we propose the original approach to impose an imperfect interface boundary condition that simulates the macroscopic behavior of a very thin elastic adhesive [20]. Often, structural adhesives have a stiffness that is comparable to the adherents' stiffnesses; in this case their mechanical behavior cannot be accurately described via a classic spring-like interface model (i.e., the continuity of stresses and discontinuity of displacements), but a hard interface condition also accounting for stress jumps is indicated more. For this reason, we assume the thin adhesive layer to be modeled by the following law of hard imperfect contact proposed in [22]:

$$[[\mathbf{u}]] = \varepsilon\left((\mathbf{K}^{33})^{-1}\left(\langle\langle\boldsymbol{\sigma}\mathbf{e}_3\rangle\rangle - \mathbf{K}^{\alpha 3}\langle\langle\mathbf{u}_{,\alpha}\rangle\rangle\right) - \langle\langle\mathbf{u}_{,3}\rangle\rangle\right) \tag{3}$$

$$[[\boldsymbol{\sigma}\,\mathbf{e}_3]] = \varepsilon\left(\left(-\mathbf{K}^{\beta\alpha}\langle\langle\mathbf{u}_{,\beta}\rangle\rangle - \mathbf{K}^{3\alpha}(\mathbf{K}^{33})^{-1}\left(\langle\langle\boldsymbol{\sigma}\mathbf{e}_3\rangle\rangle - \mathbf{K}^{\beta 3}\langle\langle\mathbf{u}_{,\beta}\rangle\rangle\right)\right)_{,\alpha}\right.$$
$$\left. -\langle\langle\boldsymbol{\sigma}_{,3}\,\mathbf{e}_3\rangle\rangle\right), \tag{4}$$

where ε is the thickness of the adhesive, and the symbols $[[(\cdot)]]$ and $\langle\langle(\cdot)\rangle\rangle$ are taken to denote the jump and the average of the quantity $(\cdot)$ across the interface separating the two adherents, respectively; the Greek indexes ($\alpha, \beta = 1, 2$) are related to the in-plane (x_1, x_2) quantities; commas denote the first derivatives, and the summation convention is used.

The transmission conditions (3) and (4) prescribe jumps in the traction $[[\boldsymbol{\sigma}\,\mathbf{e}_3]]$ and displacement $[[\mathbf{u}]]$ fields across the interface between the two adherents, thus describing the asymptotic behavior of a very thin deformable adhesive made of a general anisotropic linear-elastic material with elasticity coefficients b_{ijkl}, which are related to the matrices $\mathbf{K}^{ij}, i, j = 1, 2, 3$. If the adhesive is modeled as isotropic, with Young's modulus $\bar{E}$ and Poisson's ratio $\bar{v}$, the matrices $\mathbf{K}^{ij}$ have the form:

$$\mathbf{K}^{ii} = \frac{\bar{E}}{2(1+\bar{v})}\left(\frac{2(1-\bar{v})}{(1-2\bar{v})}\,\mathbf{e}_i\otimes\mathbf{e}_i + \mathbf{e}_j\otimes\mathbf{e}_j + \mathbf{e}_k\otimes\mathbf{e}_k\right), \quad i\neq j\neq k, \tag{5}$$

$$\mathbf{K}^{ij} = \frac{\bar{E}}{2(1+\bar{v})}\left(\mathbf{e}_i\otimes\mathbf{e}_j + \frac{2\bar{v}}{(1-2\bar{v})}\,\mathbf{e}_j\otimes\mathbf{e}_i\right), \quad j\neq i. \tag{6}$$

To take into account to the presence of the adhesive and enforce the transmission conditions (3) and (4), we propose a generalized equilibrium solution, which is obtained by modifying the displacement field (2) in the upper part of the tube-to-tube butt joint as follows:

$$\mathbf{u}(x_1, x_2, x_3) = \mathbf{u}^0(x_1, x_2, x_3) + [[u_1]]\mathbf{e}_1 + [[u_2]]\mathbf{e}_2 + [[u_3]]\mathbf{e}_3, \quad \text{for } x_3 \geq h + \varepsilon, \tag{7}$$

where the jumps $[[u_i]]$, $i = 1, 2, 3$, possibly dependent on x_1, x_2, x_3, have to be determined. In the lower part of the tube-to-tube butt joint below the adhesive, for $x_3 \leq h$, the displacement field is still given by (2).

In (7), the jumps $[[u_i]]$, $i = 1, 2, 3$, have to be chosen in order to satisfy the transmission conditions (3) and (4). In the presence of the thin deformable adhesive, assuming that the stress field is still given by (1) and substituting (1) and (5)–(7) into (3) and (4), we obtain:

$$[[u_1]] = -\varepsilon\tilde{\zeta}_G\frac{Tx_2}{GI_0},$$

$$[[u_2]] = +\varepsilon\tilde{\zeta}_G\frac{Tx_1}{GI_0},$$

$$[[u_3]] = \varepsilon\tilde{\zeta}_E\frac{F}{EA}, \tag{8}$$

with

$$\tilde{\zeta}_G = \frac{G}{\bar{G}} - 1, \tag{9}$$

$$\tilde{\zeta}_E = \frac{E}{\bar{E}}\left(1 - \frac{2\bar{v}^2}{(1-\bar{v})}\right) - \left(1 - \frac{2v\bar{v}}{(1-\bar{v})}\right), \tag{10}$$

where $\bar{G} = \bar{E}/(2(1+\bar{v}))$ is the shear modulus of the adhesive.

In view of (1), (2), (7), and (8), the generalized equilibrium solution is thus characterized by:

- The stress field (1), which is continuous across the adhesive and equilibrated by the applied loads;
- A displacement field that is discontinuous across the adhesive, which is given by (2) below the adhesive (for $0 \leq x_3 \leq h$) and by (7) above the adhesive (for $h + \varepsilon < x_3 \leq 2h + \varepsilon$);
- A strain field that is continuous across the adhesive, which is given by the symmetric part of the gradient of (2) (or (7)).

Notably, the displacement fields above and below the adhesive differ by a rotation in the (x_1, x_2)-plane, given by the jumps $[[u_1]]$ and $[[u_2]]$, and by a translation along the x_3-axis, given by the jump $[[u_3]]$. The rotation and the translation reproduce the shear and axial deformations, respectively, of a very thin adhesive under the given applied load acting on the tube-to-tube butt joint.

Finally, in the proposed generalized solution that takes the imperfect contact into account, the stress distribution is assumed to be uniform, thus neglecting the effect of stress concentrations on the behavior of the joint. This latter aspect is not addressed in the present paper.

2.3. Micro-Cracking Damaging Adhesive Model

To model a micro-cracking damaging adhesive, we consider the micromechanical homogenization approach proposed by Kachanov and Sevostianov [24,25] based on the approximation of non-interacting micro-cracks. The elastic potential in stresses (complementary energy density) of the effective medium yields the following structure for the effective modulus M, where M denotes any shear, Young's, or bulk moduli:

$$M = M_0(1 + C\rho)^{-1}, \tag{11}$$

where M_0 is the modulus of the undamaged matrix or the initial modulus of the adhesive before damage, ρ is the micro-crack density, thus representing a damage parameter, and the constant C depends on the particular modulus M that is considered and on the orientational distribution of defects. For a two-dimensional random distribution of circular voids, $C = 3$ in the Young's modulus and

$$C = \frac{(7 - 5v_0)}{2(1 - v_0^2)} \tag{12}$$

in the shear modulus, where v_0 is the Poisson ratio of the undamaged matrix [24].

2.4. Damage Evolution

Damage evolution is described as an accumulation of micro-cracks by assuming the damage parameter ρ to increase with time $t \geq 0$. The evolution of the micro-crack density for the proposed model is described by the following kinetic equation, which was proposed in [23]:

$$\eta\dot{\rho} = \left\{ \omega - \frac{1}{2} \mathbf{K}_{,\rho}(\rho) \begin{pmatrix} \langle\langle \mathbf{u}_{,1} \rangle\rangle \\ \langle\langle \mathbf{u}_{,2} \rangle\rangle \\ [[\mathbf{u}]] + \varepsilon\langle\langle \mathbf{u}_{,3} \rangle\rangle \end{pmatrix} \cdot \begin{pmatrix} \langle\langle \mathbf{u}_{,1} \rangle\rangle \\ \langle\langle \mathbf{u}_{,2} \rangle\rangle \\ [[\mathbf{u}]] + \varepsilon\langle\langle \mathbf{u}_{,3} \rangle\rangle \end{pmatrix} \right\}_+ , \tag{13}$$

where η is a positive viscosity parameter, a dot denotes time differentiation, ω is a strictly negative parameter, $\mathbf{u}$ is the generalized displacement field defined by (7) above the adhesive and by (2) below it, $\{\cdot\}_+$ denotes the positive part, and $\mathbf{K}_{,\rho}(\rho)$ indicates the component-wise derivative of the stiffness tensor

$$\mathbf{K}(\rho) = \begin{pmatrix} \varepsilon\mathbf{K}^{11} & \varepsilon\mathbf{K}^{21} & \mathbf{K}^{31} \\ \varepsilon\mathbf{K}^{12} & \varepsilon\mathbf{K}^{22} & \mathbf{K}^{32} \\ \mathbf{K}^{13} & \mathbf{K}^{23} & \frac{1}{\varepsilon}\mathbf{K}^{33} \end{pmatrix} \tag{14}$$

with respect to ρ. Note that $\mathbf{K}_{,\rho}(\rho)$ also depends on the adhesive layer thickness ε.

The kinetic Equation (13) is a first-order ODE in the unknown damage evolution function $\rho = \rho(t)$ to be solved for the initial condition $\rho(0) = \rho_0$. It is important to emphasize that (13) is directly related to the intrinsic mechanical and damage properties of the adhesive layer. In detail, η is a damage viscosity that influences the velocity of the damage evolution, and ω is an energy threshold, which is similar to the energy of adhesion of polymers [26], after which the damage evolution starts at the adhesive layer.

2.5. Stress–Strain Response

The aim here is to find the stress–strain response of the adhesive in the tube-to-tube butt joint subjected to a combined tensile–torsion loading. The tensile stress σ and shear stress τ in the adhesive layer are calculated as:

$$\sigma = \frac{F}{A}, \quad \tau = \frac{T}{I_0}R, \tag{15}$$

where R is the outer radius of the joint. The tensile strain ϵ and shear strain γ of the adhesive are given by:

$$\epsilon = \frac{[[u_3]]}{\varepsilon}, \quad \gamma = \frac{\sqrt{[[u_1]]^2 + [[u_2]]^2}}{\varepsilon}, \tag{16}$$

respectively, where $[[u_3]]$ is the axial displacement of the adhesive and the square root is the circumferential displacement at the outer diameter of the adhesive. Substituting (8) into (16), the normalized tensile stress–tensile strain and shear stress–shear strain are found as follows:

$$\sigma/E = \xi_E^{-1}\epsilon, \quad \tau/G = \xi_G^{-1}\gamma, \tag{17}$$

where ξ_G, ξ_E are given by (9) and (10), respectively. Note that in (9) and (10), the moduli $\bar{E}$ and $\bar{G}$ depend upon the micro-crack density ρ through the KS model (cf. (11)) as follows:

$$\bar{E} = E_0(1 + C_E\rho)^{-1}, \quad \bar{G} = G_0(1 + C_G\rho)^{-1}, \tag{18}$$

where $G_0 = E_0/(2(\nu_0 + 1))$ is the initial shear modulus of the adhesive. The damage parameter ρ evolves via the kinetic Equation (13). By substituting (2), (5)–(7), (14), and (18) into (13) and simplifying, we obtain the following evolution problem for the damage parameter $\rho = \rho(t)$:

$$\begin{cases} \eta\dot{\rho} = \{\omega + \mathcal{F}(\rho, F) + \mathcal{T}(\rho, T)\}_+, \\ \rho(0) = \rho_0, \end{cases} \tag{19}$$

with

$$\mathcal{F}(\rho, F) = \left(C_1 + \frac{C_2}{(C_3 + C_4\rho)^2}\right)\left(\frac{\varepsilon F^2}{2A^2E^2E_0}\right), \tag{20}$$

$$\mathcal{T}(\rho, T) = \frac{\varepsilon C_G T^2 R^2}{2G_0 I_0^2}. \tag{21}$$

The constants C_i, $i = 1, 2, 3, 4$, reported in Appendix A, depend on the elasticity coefficients of the adherents E, ν, on the initial elasticity coefficients of the adhesive E_0, ν_0, and on the constants C_E and C_G. Note that tensile and torsion loads are decoupled in (19). Finally, since we are simulating force-controlled tests, the use of (15) allows us to eliminate the tensile load F and the torque T in favor of the control variables $\sigma(t) = \dot{\sigma}t$ and $\tau(t) = \dot{\tau}t$, where $\dot{\sigma}$ and $\dot{\tau}$ are the tensile and shear strain rates, respectively.

For pure torsion, i.e., for $F = 0$ and $T \neq 0$, the evolution problem (19) admits the simple solution:

$$\rho(t) = \begin{cases} \rho_0, & 0 \leq t \leq t_0 \\ \rho_0 + \frac{\omega}{\eta}(t - t_0) + \frac{\varepsilon C_G \dot{t}}{6 G_0 \eta}(t^3 - t_0^3), & t > t_0 \end{cases} \tag{22}$$

where t_0 is the instant at which damage evolution begins:

$$t_0 = \frac{1}{\dot{t}}\sqrt{-\frac{2\omega G_0}{\varepsilon C_G}}. \tag{23}$$

Substituting (22) and (9) into the second of (17), the shear strain–stress response of the adhesive is obtained:

$$\gamma = \begin{cases} a\tau, & 0 \leq \tau \leq \tau_0 \\ a\tau + b\tau(\tau - \tau_0)^2(\tau + 2\tau_0), & \tau > \tau_0 \end{cases} \tag{24}$$

with

$$\tau_0 = \dot{\tau} t_0 = \sqrt{-\frac{2\omega G_0}{\varepsilon C_G}}, \tag{25}$$

$$a = \frac{1}{G_0} - \frac{1}{G} + \rho_0\frac{C_G}{G_0}, \tag{26}$$

$$b = \frac{\varepsilon C_G^2}{6\eta\dot{t}G_0^2}. \tag{27}$$

For a pure tensile load, i.e., for $F \neq 0$ and $T = 0$, or for a combined tensile–torsion loading, a general closed-form solution of the evolution problem (19) is not available. However, it is possible to obtain a closed-form solution before damage initiation. Indeed, in view of the positivity of the constants C_1 and C_2 in (20) (cf. the Appendix A), inspection of (19) indicates that the instant t_0 of damage initiation for a generic combination of tensile and torsion loads takes the form:

$$t_0 = \sqrt{-\frac{\omega}{\left[C_1 + \frac{C_2}{(C_3+C_4\rho)^2}\right]\frac{\varepsilon\dot{\sigma}^2}{2E^2 E_0} + \frac{\varepsilon C_G\dot{t}^2}{2G_0}}}. \tag{28}$$

Note that for pure torsion ($\dot{\sigma} = 0$ and $\dot{t} \neq 0$), (28) reduces to (23).

For $t \leq t_0$, the shear stress–strain response is still given by the linear part in (24), while the tensile stress–strain response takes the following linear form:

$$\sigma = \frac{(C_5 + C_6\rho_0)}{(C_7 + C_8\rho_0 + C_9\rho_0^2)}\epsilon \tag{29}$$

where the constants C_i, $i = 5, 6, 7, 8, 9$ are given in Appendix A.

3. Numerical Implementation

The numerical simulations for the pure tensile and for a combined tensile–torsion loading condition were carried out by numerically solving the differential problem (19) using the commercial software *Mathematica* [27]. For pure torsion loading, the closed-form solution (24) was used. Tables 1 and 2 show the geometrical and material parameters of the joint specimen of the experimental study by Murakami and coworkers [9] that were chosen to compare with those of the proposed model as an original validation. In [9], the adherents were two S45C carbon steel cylinders joined by a one-component epoxy adhesive (XA7416, 3M Japan Ltd., Tokyo, Japan).

Table 1. Geometrical parameters of the joint specimen [9].

Quantity	Symbol	Value	Unit
Outer diameter	D	26.0	mm
Inner diameter	d	20.0	mm
Adhesive thickness	ε	0.3	mm

Table 2. Mechanical properties of the joint materials [9].

Quantity	Symbol	Value	Unit
Adhesive Young's modulus	E_0	4.53	GPa
Adhesive Poisson's ratio	ν_0	0.36	–
Adherents' Young's modulus	E	200.00	GPa
Adherents' Poisson's ratio	ν	0.30	–

The micromechanical parameters C_E and C_G in (18) were chosen to be equal to 3.00 and to 2.98, respectively, the latter value being estimated using (12). The other micromechanical parameters, i.e., the initial value of the damage parameter ρ_0, the viscosity parameter η, and the energy threshold ω, will be identified in the next subsections starting from the experimental data from [9].

According to [9], two different stress rates were considered in the numerical analyses: 6.67×10^{-2} MPa/s for the quasi-static (QS) condition and 1.00×10^3 MPa/s for the high-rate (HR) condition.

For pure tensile and torsion tests, the simulations were stopped at failure, i.e., when the stress reached the tensile and shear limit strengths, respectively. For a combined tensile–torsion test, the tensile and shear stresses were related using a loading angle θ:

$$\theta = \arctan\left(\frac{\tau}{\sigma}\right). \tag{30}$$

The tensile and shear strengths estimated experimentally in [9] for some values of the loading angle are reported in Table 3.

Table 3. Experimentally estimated tensile and torsional (shear) strengths of butt-joint specimens studied in [9].

Loading Rate	Loading Angle [deg.]	Tensile Strength [MPa]	Shear Strength [MPa]	Failure Strain in Tension [%]	Failure Strain in Shear [%]
QS	0	61.8	–	4.0	–
HR	0	90.0	–	6.1	–
QS	90	–	53.2	–	37.0
HR	90	–	70.0	–	32.0
QS	18.0	61.0	19.9	2.15	2.85
HR	15.5	92.1	25.5	3.55	3.69

4. Results and Discussion

In what follows, a first validation of the proposed model is illustrated. Experimental data obtained in [9] were chosen for comparison in order to highlight the capacity of our model to reproduce the stress–strain behavior of adhesive layers under both pure and combined loads for quasi-static and high-rate loading conditions.

4.1. Simulation of Pure Tensile Tests

Figure 2a shows the stress–strain curves of the adhesive layer in a pure tensile load obtained by the proposed model for quasi-static (gray) and high-rate (black) loading (solid lines), compared with the experimental curves by [9] (dashed lines). The experimental data were extracted from Figure 9 in [9] by using the free online software WebPlotDigitizer [28].

In view of the modeling approach proposed in the present paper, the experimental curves in Figure 2a can be interpreted as a linear stress–strain response in analogy with a brittle damage behavior, for which the accumulated damage slightly differs from the initial damage ρ_0. Accordingly, by fitting the experimental curves in Figure 2a into Mathematica by using a linear model, we obtained the following slopes: 16.97×10^2 MPa for the quasi-static case and 17.83×10^2 MPa for the high-rate case. From (29), the initial value of the damage parameter, ρ_0, was calculated as follows:

- $\rho_0 = 1.14$ for the QS case;
- $\rho_0 = 1.07$ for the HR case.

These two values are very close. Accordingly to our model, this would indicate a very similar micro-crack density in the samples tested in [9], both in quasi-static and high-rate loading conditions.

Thus, the proposed model is fully able to reproduce the stress–strain behavior under pure tensile loading. Moreover, it is able to catch the influence of the loading rate that was found experimentally in [9], meaning that the tensile strength is higher for high-rate load. It is important to emphasize that this brittle damage behavior of structural adhesive layers under tensile loads has been found in other experimental work, such as that of [11,29].

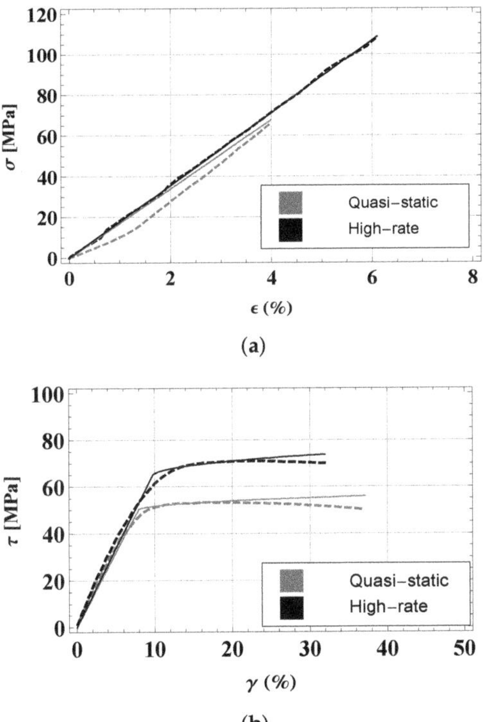

Figure 2. Stress–strain curves in pure loading conditions: (**a**) Stress–strain curves of the adhesive layer under a pure tensile load obtained with the proposed model for quasi-static and high-rate loading (solid lines) compared with experimental curves by [9] (dashed lines). (**b**) Stress–strain curves of the adhesive layer under a pure torsion load obtained with the proposed model for quasi-static and high-rate loading (solid lines) compared with experimental curves by [9] (dashed lines).

4.2. Simulation of Pure Torsion Tests

Figure 2b shows the stress–strain curves of the adhesive layer under a pure torsional load obtained with the proposed model for quasi-static (gray) and high-rate (black) loading (solid lines) compared with the experimental curves by [9] (dashed lines). The experimental data were extracted from Figure 10 in [9] by using the free online software WebPlotDigitizer [28].

By fitting the experimental data into Mathematica by using a nonlinear model based on (24), the following values were obtained for the parameters a, b, and τ_0 :

- $a = 1.58 \times 10^{-3}$ MPa^{-1}, $b = 1.05 \times 10^{-6}$ MPa^{-4}, $\tau_0 = 50.25$ MPa for the QS case;
- $a = 1.50 \times 10^{-3}$ MPa^{-1}, $b = 1.78 \times 10^{-7}$ MPa^{-4}, $\tau_0 = 64.97$ MPa for the HR case.

Substituting these data into (25)–(27), the following values of the parameters ρ_0, η, and ω were calculated:

- $\rho_0 = 0.55$, $\eta = 2296.70$ Ns/m, $\omega = -679.34$ N/m for the QS case;
- $\rho_0 = 0.51$, $\eta = 0.90$ Ns/m, $\omega = -1135.64$ N/m for the HR case.

The theoretical curves were stopped at the failure strains—37% for the QS case and 32% for the HR case. As a result, the relative errors between the experimental values (reported in Table 3 for pure torsion, i.e., for $\theta = 90°$) and theoretical failure strengths are equal to 4.8% in the QS case and 5.5% in the HR case. On the other hand, after damage initiation, i.e., for $\tau > \tau_0$, both experimental curves exhibited a softening behavior, which cannot be reproduced by the force-controlled theoretical model.

Both the experimental and theoretical stress–strain curves under pure torsion are typical of a ductile damage behavior of the structural adhesive. The proposed model is thus clearly able to accurately reproduce this kind of behavior in terms of both yielding and failure stresses. A ductile damage behavior of structural adhesives in tube-to-tube butt joints is commonly found in torsion experiments (see, for example, the work by Kosmann and coworkers [10]). In addition, in the case of pure torsional tests, the yielding and the failure stresses for the high-rate load were higher than those for the quasi-static load (of almost 34% according to [9]), and the proposed model was able to catch this experimental finding.

4.3. Simulation of Combined Tensile–Torsion Tests

Using the values of the micromechanical parameters ρ_0, η, and ω identified in the previous subsections for the pure loading cases, it was possible to solve the damage evolution problem (19) and plot the corresponding stress–strain diagrams for a combined tensile–torsion load. In particular, the experimental combined loading conditions from [9] for $\theta = 18.0°$ under a QS load and for $\theta = 15.5°$ under an HR load (see Table 3) were selected to be simulated via the proposed analytical model. The limit strengths in the tensile and shear conditions were set up to stop simulations in the stress-controlled mode, and the failure strains in tension and shear were obtained. The simulated stress–strain curves are plotted in Figure 3; the tensile part is shown in Figure 3a and the torsional part in Figure 3b. The corresponding experimental stress–strain curves are not available in [9], so a direct comparison was not possible. However, it was possible to compare the values of the strains at failure. For the QS loading condition ($\theta = 18.0°$), the simulations gave failure strains of 2.14% in tension and 3.13% in torsion, providing acceptable relative errors (0.46% and 9.82%) when compared to the failure strains estimated experimentally (2.15% and 2.85%). For the HR loading condition ($\theta = 15.5°$), the simulations gave failure strains of 3.08% in tension and 3.83% in torsion, again providing acceptable relative errors (13.24% and 3.80%) when compared to the failure strains estimated experimentally (3.55% and 3.69%).

Finally, the simulated stress–strain curves for the combined tensile–torsion load exhibited a brittle damage behavior. Unfortunately, it was not possible to find experimental curves in the literature in order to make a comparison, so this and related aspects will be investigated in further work.

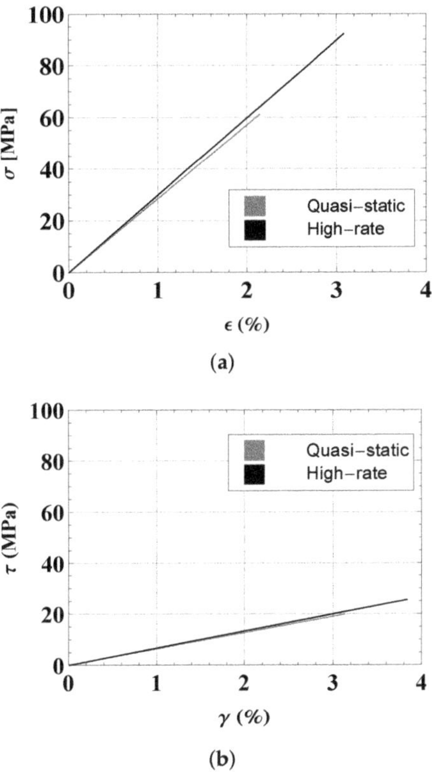

Figure 3. Stress–strain curves in combined tensile–torsion loading conditions: (**a**) Tensile stress–strain curves of the adhesive layer obtained with the proposed model for quasi-static and high-rate loading. (**b**) Torsion stress–strain curves of the adhesive layer obtained with the proposed model for quasi-static and high-rate loading.

5. Summary

The behavior of thin adhesive layers in butt joints under combined tensile and torsion loads was modeled by using an imperfect interface approach that merged continuum damage mechanics and asymptotic homogenization. The proposed approach took micro-cracking damage evolution into account, resulting in a ductile stress–strain behavior of the adhesive for the pure torsional tests and in a brittle stress–strain behavior for the pure tensile and combined tensile–torsion tests. In the case of pure torsion (ductile damage behavior), a closed-form solutions was proposed. In the case of brittle damage behavior (pure tensile and combined tensile–torsion loads), a closed-form solution was calculated in the linear stress–strain domain. The comparisons with the experimental data from [9] gave satisfying results in terms of the failure strains for pure and combined loads in both QS and HR conditions. In all cases, the relative errors between the experimental and simulated failure strains were found to be less than 14%.

The proposed model has some main limitations. First, stresses in the adhesive layer are supposed to be uniformly distributed. This is not realistic, particularly at the boundaries between the adhesive and adherents, where stress concentrations are known to occur [11]; therefore, the effect of the stress concentration on tensile strength is not discussed in this paper. Next, for the sake of simplicity, the adhesive thickness was assumed to be constant and uniform in the whole layer, and perfect thickness uniformity is almost impossible to achieve in real applications. Nevertheless, it is possible to easily generalize the analytical model by accounting for a smooth–rough interface (cf. [19]). Lastly, the viscoplasticity and viscoelasticity that are typical of structural adhesives were not considered in the proposed model. These aspects could be the object of further work.

Despite these limitations, the model is able to accurately reproduce experimental stress–strain behavior for both brittle and ductile damages. Future studies will focus on an experimental protocol for the identification of the model parameters.

Author Contributions: Conceptualization, all authors; methodology, M.L.R. and R.R.; software, M.L.R. and R.R.; validation, all authors; formal analysis, R.R.; investigation, all authors; resources, R.R.; data curation, M.L.R.; writing—original draft preparation, M.L.R. and R.R.; writing—review and editing, M.L.R. and R.R.; visualization, M.L.R. and R.R.; supervision, F.L. and R.R.; project administration, F.L.; funding acquisition, F.L. All authors have read and agreed to the published version of the manuscript.

Funding: This research was funded by the University of Ferrara through FAR grants (2020 and 2021, R.R.).

Institutional Review Board Statement: Not applicable.

Informed Consent Statement: Not applicable.

Data Availability Statement: No new data were created or analyzed in this study. Data sharing is not applicable to this article.

Conflicts of Interest: The authors declare no conflict of interest.

Appendix A

Constants in (20):

$$C_1 = \frac{C_G E^2 (v_0 + 1)(2C_G(v_0 + 1) - 3C_E)}{C_G(1 + v_0) - 2C_E}, \tag{A1}$$

$$C_2 = \frac{2(C_G(v_0 + 1)(E - E_0 v) - C_E(Ev_0 + E - 2E_0 v))^2}{2C_E - C_G(v_0 + 1)}, \tag{A2}$$

$$C_3 = 1 - v_0, \tag{A3}$$

$$C_4 = 2C_E - C_G(1 + v_0). \tag{A4}$$

In the case of a bi-dimensional circular defect, one has $C_E = 3$ and $C_G = (7 - 5v_0)/(2(1 - v_0^2))$, and the constants C_1, C_2 and C_4 simplify to:

$$C_1 = \frac{2E^2(1 - 2v_0)(7 - 5v_0)}{(1 - v_0)(5 - 7v_0)}, \tag{A5}$$

$$C_2 = \frac{(E(v_0(6v_0 - 5) + 1) + E_0 v(5 - 7v_0))^2}{(1 - v_0)(5 - 7v_0)} \tag{A6}$$

$$C_4 = \frac{1}{(1 - v_0)} - \frac{7}{2}. \tag{A7}$$

Assuming $-1 < v_0 < 1/2$, the constants C_1, C_2 and C_3 are positive, and C_4 is negative.

Constants in (29):

$$C_5 = -\frac{EE_0^2(\nu_0 - 1)}{2(\nu_0 + 1)^2},\tag{A8}$$

$$C_6 = -\frac{EE_0^2(-2C_E + C_G\nu_0 + C_G)}{2(\nu_0 + 1)^2},\tag{A9}$$

$$C_7 = \frac{E_0\big(E_0(2\nu\nu_0 + \nu_0 - 1) - E(2\nu_0^2 + \nu_0 - 1)\big)}{2(\nu_0 + 1)^2},\tag{A10}$$

$$C_8 = \frac{E_0\big(C_E(3E(\nu_0 + 1) - 2E_0(\nu + 1)) - C_G(\nu_0 + 1)(4E\nu_0 + E - E_0(2\nu + 1))\big)}{2(\nu_0 + 1)^2},\tag{A11}$$

$$C_9 = C_G E\left(\frac{3C_E E_0}{2(\nu_0 + 1)} - C_G E_0\right).\tag{A12}$$

For $C_E = 3$ and $C_G = (7 - 5\nu_0)/(2(1 - \nu_0^2))$, the constants C_6, C_8 and C_9 specialize as

$$C_6 = \frac{EE_0^2(7\nu_0 - 5)}{4(\nu_0 - 1)(\nu_0 + 1)^2},\tag{A13}$$

$$C_8 = -\frac{E_0\big(E(\nu_0 - 11)(2\nu_0 - 1) + E_0(2\nu(\nu_0 + 1) + 7\nu_0 - 5)\big)}{4(\nu_0 - 1)(\nu_0 + 1)^2},\tag{A14}$$

$$C_9 = \frac{EE_0(2\nu_0 - 1)(5\nu_0 - 7)}{2(\nu_0^2 - 1)^2}.\tag{A15}$$

References

1. Higgins, A. Adhesive bonding of aircraft structures. *Int. J. Adhes. Adhes.* **2000**, *20*, 367–376.
2. Maurel-Pantel, A.; Lamberti, M.; Raffa, M.L.; Suarez, C.; Ascione, F.; Lebon, F. Modelling of a GFRP adhesive connection by an imperfect soft interface model with initial damage. *Compos. Struct.* **2020**, *239*, 112034.
3. Kawecki, B.; Podgórski, J. The Effect of Glue Cohesive Stiffness on the Elastic Performance of Bent Wood–CFRP Beams. *Materials* **2020**, *13*, 5075.
4. Cavezza, F.; Boehm, M.; Terryn, H.; Hauffman, T. A Review on Adhesively Bonded Aluminium Joints in the Automotive Industry. *Metals* **2020**, *10*, 730.
5. Sokolowski, G.; Krasowski, M.; Szczesio-Wlodarczyk, A.; Konieczny, B.; Sokolowski, J.; Bociong, K. The Influence of Cement Layer Thickness on the Stress State of Metal Inlay Restorations—Photoelastic Analysis. *Materials* **2021**, *14*, 599.
6. Ramalho, L.D.C.; Campilho, R.D.S.G.; Belinha, J.; Da Silva, L.F.M. Static strength prediction of adhesive joints: A review. *Int. J. Adhes. Adhes.* **2020**, *96*, 102451.
7. Rucka, M.; Wojtczak, E.; Lachowicz, J. Damage imaging in Lamb wave-based inspection of adhesive joints. *Appl. Sci.* **2018**, *8*, 522.
8. Chai, H. Observation of deformation and damage at the tip of cracks in adhesive bonds loaded in shear and assessment of a criterion for fracture. *Int. J. Fract.* **1993**, *60*, 311–326.
9. Murakami, S.; Sekiguchi, Y.; Sato, C.; Yokoi, E.; Furusawa, T. Strength of cylindrical butt joints bonded with epoxy adhesives under combined static or high-rate loading. *Int. J. Adhes. Adhes.* **2016**, *67*, 86–93.
10. Kosmann, J.; Klapp, O.; Holzhüter, D.; Schollerer, M.J.; Fiedler, A.; Nagel, C.; Hühne, C. Measurement of epoxy film adhesive properties in torsion and tension using tubular butt joints. *Int. J. Adhes. Adhes.* **2018**, *83*, 50–58.
11. Spaggiari, A.; Castagnetti, D.; Dragoni, E. Mixed-mode strength of thin adhesive films: Experimental characterization through a tubular specimen with reduced edge effect. *J. Adhes.* **2013**, *89*, 660–675.
12. Zhang, J.; Wang, J.; Yuan, Z.; Jia, H. Effect of the cohesive law shape on the modelling of adhesive joints bonded with brittle and ductile adhesives. *Int. J. Adhes. Adhes.* **2018**, *85*, 37–43.
13. Campilho, R.D.; Banea, M.D.; Neto, J.A.B.P.; da Silva, L.F. Modelling adhesive joints with cohesive zone models: Effect of the cohesive law shape of the adhesive layer. *Int. J. Adhes. Adhes.* **2013**, *44*, 48–56.
14. Hu, C.; Huang, G.; Li, C. Experimental and Numerical Study of Low-Velocity Impact and Tensile after Impact for CFRP Laminates Single-Lap Joints Adhesively Bonded Structure. *Materials* **2021**, *14*, 1016.
15. Yu, G.; Chen, X.; Zhang, B.; Pan, K.; Yang, L. Tensile-Shear Mechanical Behaviors of Friction Stir Spot Weld and Adhesive Hybrid Joint: Experimental and Numerical Study. *Metals* **2020**, *10*, 1028.
16. Liljedahl, C.D.M.; Crocombe, A.D.; Wahab, M.A.; Ashcroft, I.A. Modelling the environmental degradation of adhesively bonded aluminium and composite joints using a CZM approach. *Int. J. Adhes. Adhes.* **2007**, *27*, 505–518.
17. Pirondi, A.; Moroni, F. Improvement of a Cohesive Zone Model for Fatigue Delamination Rate Simulation. *Materials* **2019**, *12*, 181.
18. Lebon, F.; Dumont, S.; Rizzoni, R.; López-Realpozo, J.C.; Guinovart-Díaz, R.; Rodríguez-Ramos, R.; Bravo-Castillero, J.; Sabina, F.J. Soft and hard anisotropic interface in composite materials. *Compos. Part Eng.* **2016**, *90*, 58–68.

19. Dumont, S.; Lebon, F.; Raffa, M.L.; Rizzoni, R. Towards nonlinear imperfect interface models including microcracks and smooth roughness. *Ann. Solid Struct. Mech.* **2017**, *9*, 13–27.
20. Benveniste, Y.; Miloh, T. Imperfect soft and stiff interfaces in two-dimensional elasticity. *Mech. Mater.* **2001**, *33*, 309–323.
21. Furtsev, A.; Rudoy, E. Variational approach to modeling soft and stiff interfaces in the Kirchhoff-Love theory of plates. *Int. J. Solids Struct.* **2020**, *202*, 562–574.
22. Rizzoni, R.; Dumont, S.; Lebon, F.; Sacco, S. Higher order model for soft and hard interfaces. *Int. J. Solids Struct.* **2014**, *51*, 4137–4148.
23. Raffa, M.L.; Lebon, F.; Rizzoni, R. A micromechanical model of a hard interface with micro-cracking damage. In Preparation.
24. Kachanov, M. Elastic solids with many cracks and related problems. *Adv. Appl. Mech.* **1994**, *30*, 259–445.
25. Kachanov, M.; Sevostianov, I. *Micromechanics of Materials, with Applications*; Springer: Cham, Switzerland, 2018; Volume 249.
26. Weng, C.; Yang, J.; Yang, D.; Jiang, B. Molecular Dynamics Study on the Deformation Behaviors of Nanostructures in the Demolding Process of Micro-Injection Molding. *Polymers* **2019**, *11*, 470.
27. Wolfram Research, Inc. *Mathematica*; Version 12.2; Wolfram Research, Inc.: Champaign, IL, USA, 2020.
28. Ankit Rohatgi. WebPlotDigitizer. Version: 4.4. November 2020. Available online: https://automeris.io/WebPlotDigitizer (accessed on December 2020).
29. Costa-Mattos, H.S.; Monteiro, A.H.; Sampaio, E.M. Modelling the strength of bonded butt-joints. *Compos. Part B Eng.* **2010**, *41*, 654–662.

Article

Interface Models in Coupled Thermoelasticity

Michele Serpilli [1,*] , Serge Dumont [2] , Raffaella Rizzoni [3] and Frédéric Lebon [4]

1   Department of Civil and Building Engineering, and Architecture, Università Politecnica delle Marche, 60121 Ancona, Italy
2   IMAG CNRS UMR 5149, University of Nîmes, 30000 Nîmes, France; serge.dumont@unimes.fr
3   Department of Engineering, University of Ferrara, 44122 Ferrara, Italy; raffaella.rizzoni@unife.it
4   CNRS, Centrale Marseille, Laboratoire de Mécanique et d'Acoustique, Aix-Marseille University, 13453 Marseille, France; lebon@lma.cnrs-mrs.fr
*   Correspondence: m.serpilli@univpm.it

Abstract: This work proposes new interface conditions between the layers of a three-dimensional composite structure in the framework of coupled thermoelasticity. More precisely, the mechanical behavior of two linear isotropic thermoelastic solids, bonded together by a thin layer, constituted of a linear isotropic thermoelastic material, is studied by means of an asymptotic analysis. After defining a small parameter ε, which tends to zero, associated with the thickness and constitutive coefficients of the intermediate layer, two different limit models and their associated limit problems, the so-called *soft* and *hard* thermoelastic interface models, are characterized. The asymptotic expansion method is reviewed by taking into account the effect of higher-order terms and defining a generalized thermoelastic interface law which comprises the above aforementioned models, as presented previously. A numerical example is presented to show the efficiency of the proposed methodology, based on a finite element approach developed previously.

Keywords: interfaces; asymptotic analysis; coupled thermoelasticity

---

Citation: Serpilli, M.; Dumont, S.; Rizzoni, R.; Lebon, F. Interface Models in Coupled Thermoelasticity. *Technologies* **2021**, *9*, 17. https://doi.org/10.3390/technologies9010017

Received: 12 February 2021
Accepted: 2 March 2021
Published: 4 March 2021

Publisher's Note: MDPI stays neutral with regard to jurisdictional claims in published maps and institutional affiliations.

1. Introduction

The use of composite structures, obtained by bonding together simpler structural members, has spread in all fields of engineering in the last decades. On the one hand, the structural assembly presents a significant improvement of the mechanical properties and an enhancement of its performances. On the other hand, the bonded joints among the composite constituents may cause a jump of the physical fields at the interface level and radically modify the global mechanical response. Thus, the correct modeling of composite interfaces is crucial in the understanding and design of complex structures.

From a theoretical point of view, the bonded region is considered as a thin interphase between two adjacent parts. By letting the thickness of this layer tend to zero, the interphase is reduced into a two-dimensional surface, called imperfect interface, where ad-hoc transmission conditions in terms of the representative physical fields are prescribed. The contact laws can be derived by means of classical variational tools and more refined mathematical techniques, in different physical frameworks, involving uncoupled (thermal conduction and elasticity) and coupled (piezoelectricity and multiphysics) phenomena.

Concerning the thermal (or electrical) conduction case, two main interface laws have been formulated: the lowly-conducting (LC) or Kapitza's model and highly-conducting (HC) model. The LC model provides a discontinuity of the temperature field (electric potential) and a continuity of the normal heat flow (electric displacement) across the interface (see, e.g., [3–5]). The HC model gives rise to two-dimensional Young–Laplace equation, defined on the interface, depending on the jump of the normal heat flow (electric displacement) and maintaining the temperature (electric potential) continuous (see, e.g., [6,7]). A unifying approach of a general imperfect interface model, involving the concurrent jump of both the temperature field and the normal heat flow, recovering both the LC and HC models, was proposed by [8,9]. Concerning the linear elastic case, three types of imperfect interfaces have been proposed: the spring-layer interface model (SL) (soft interface), the coherent interface (CI) (rigid interface), and the general imperfect interface. The SL models considers that the traction vector is continuous across the interface, while the displacement presents a jump linearly proportional to the traction vector (see, e.g., [10,11]). The CI model has been developed for continuum theories with surface effects and nano-sized materials (see, e.g., [12–14]): the traction vector suffers a jump, while the displacement field is continuous across the interface. Finally, in the general imperfect model, both the displacement and normal traction fields are discontinuous across the interface [15,16].

The asymptotic expansions method and convergence approaches represent mathematical tools, usually employed in the derivation and justification of classical thin structures and layered plates [17–21]. These methodologies are based on the behavior of the problem solution, when a small parameter ε, related to the thickness of the interphase, tends to zero. Considering that the material properties of the intermediate layer depend on ε^p, different limit behaviors can be derived by means of the asymptotic analysis: for $p = 1$, an SL interface model can be recovered (see, e.g., [22,23]); and, for $p = -1$, the CI interface model is mathematically justified by means of strong convergence arguments in [24,25]. Within the framework of a higher-order theory, assuming the interphase elastic constants are independent of the small thickness ($p = 0$), the asymptotic analysis yields to a general stiff imperfect interface condition, prescribing both the jumps of the displacement and traction vector fields and recovering as a particular case the perfect contact conditions at the zeroth-order [26–30]. The above transmission conditions have been generalized by considering some multiphysics and multifield couplings, such as in piezoelectricity and magneto-electro-thermo-elasticity [31,32], poroelasticity [33], and micropolar elasticity [34].

The goal of the present work is to identify the interface limit models of a composite constituted by a thin thermoelastic layer surrounded by two thermoelastic bodies in the framework of dynamic coupled thermoelasticity. Different situations are analyzed by varying the stiffnesses ratios between the middle layer and the adherents: namely, the *soft* thermoelastic lowly conducting interface, where the intermediate material coefficients have the order of magnitude ε with respect to those of the surrounding bodies, and the *hard* thermoelastic moderately conducting interface, where the constitutive parameters have the same order of magnitude. Following the ideas of [1], a generalized interface law is derived, comprising the aforementioned behaviors. A numerical investigation was performed in the framework of the finite element method (FEM), employing the approach developed in [2] for multiphysics problems, in order to assess the validity of the asymptotic models. Convergence results and a comparison between the full 3D model and the generalized interface problem are given.

2. Position of the Problem

In the sequel, Greek indices range in the set $\{1, 2\}$, Latin indices range in the set $\{1, 2, 3\}$, and the Einstein's summation convention with respect to the repeated indices is adopted. Let us consider a three-dimensional Euclidian space identified by $\mathbb{R}^3$ and such that the three vectors $_i$ form an orthonormal basis. Let $\mathbb{M}^n$ be the space of $n \times n$ square matrices. We introduce the following notations for the inner products: $\mathbf{a} \cdot \mathbf{b} := a_i b_i$, for all vectors $\mathbf{a} = (a_i)$ and $\mathbf{b} = (b_i)$ in $\mathbb{R}^3$ and $\mathbf{A} : \mathbf{B} := A_{ij} B_{ij}$, for all $\mathbf{A} = (A_{ij})$ and $\mathbf{B} = (B_{ij})$ in $\mathbb{M}^3$.

Let us define a small parameter $0 < \varepsilon < 1$. We consider the assembly constituted of two solids $\Omega^\varepsilon_\pm \subset \mathbb{R}^3$, called the adherents, bonded together by an intermediate thin layer $B^\varepsilon := S \times (-\frac{\varepsilon}{2}, \frac{\varepsilon}{2})$ of thickness ε, called the adhesive, with cross-section $S \subset \mathbb{R}^2$. In the following, B^ε and S are called interphase and interface, respectively. Let $S^\varepsilon_\pm$ be the plane interfaces between the interphase and the adherents and let $\Omega^\varepsilon := \Omega^\varepsilon_+ \cup B^\varepsilon \cup \Omega^\varepsilon_-$ denote the composite system comprising the interphase and the adherents (cf. Figure 1a).

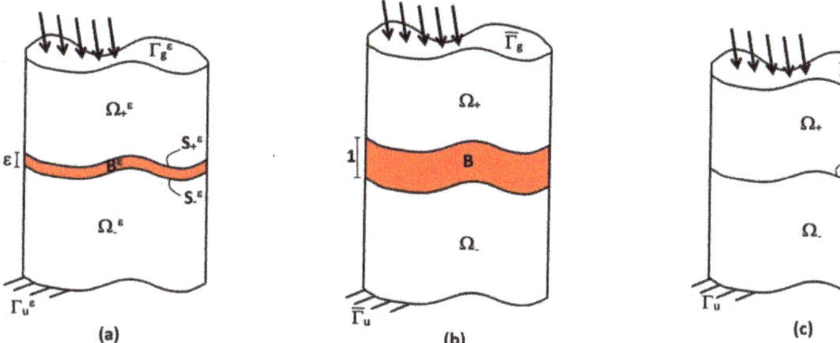

Figure 1. Initial (**a**); rescaled (**b**); and limit (**c**) configurations of the composite.

Let us assume that $\Omega^\varepsilon_\pm$ and B^ε are constituted by three homogeneous linear isotropic thermoelastic materials, whose constitutive laws are defined as follows:

$$\begin{cases} \sigma^\varepsilon_{ij} = \lambda^\varepsilon e^\varepsilon_{pp}\delta_{ij} + 2\mu^\varepsilon e^\varepsilon_{ij} - \beta^\varepsilon\theta^\varepsilon\delta_{ij}, \\ S^\varepsilon = c^\varepsilon_v\theta^\varepsilon + \beta^\varepsilon e^\varepsilon_{pp}, \\ q^\varepsilon_i = -k^\varepsilon\theta^\varepsilon_{,i}, \end{cases} \tag{1}$$

where $\sigma^\varepsilon = (\sigma^\varepsilon_{ij})$ is the Cauchy stress tensor, associated with the linearized strain tensor $e^\varepsilon = (e^\varepsilon_{ij}) := \frac{1}{2}(u^\varepsilon_{i,j} + u^\varepsilon_{j,i})$, S^ε represents the thermodynamic entropy and $q^\varepsilon = (q^\varepsilon_i)$ is the heat flow field. Constants λ^ε, μ^ε, β^ε, c^ε_v, and k^ε represent the Lamé's constants, the thermal stress coefficient, the calorific capacity, and the thermal conductivity, respectively.

The thermoelastic state is defined by the couple $s^\varepsilon := (u^\varepsilon, \theta^\varepsilon)$, where u^ε and θ^ε are the displacement field and variation of temperature, respectively. The thermoelastic composite is subject to body forces $f^\varepsilon = (f^\varepsilon_i) : \Omega^\varepsilon_\pm \times (0,T) \to \mathbb{R}^3$ and heat source $h^\varepsilon : \Omega^\varepsilon_\pm \times (0,T) \to \mathbb{R}$, applied on the top and bottom bodies, while all thermo-mechanical loadings and inertia forces are neglected in the intermediate layer B^ε. The thermoelastic state s^ε verifies the following coupled thermoelasticity system:

$$\begin{cases} \rho^\varepsilon\ddot{u}^\varepsilon - \operatorname{div}^\varepsilon\sigma^\varepsilon = f^\varepsilon & \text{in } \Omega^\varepsilon_\pm \times (0,T), \\ \dot{S}^\varepsilon + \frac{1}{T_0}\operatorname{div}^\varepsilon q^\varepsilon = h^\varepsilon & \text{in } \Omega^\varepsilon_\pm \times (0,T), \end{cases} \quad \begin{cases} \operatorname{div}^\varepsilon\sigma^\varepsilon = 0 & \text{in } B^\varepsilon \times (0,T), \\ \dot{S}^\varepsilon + \frac{1}{T_0}\operatorname{div}^\varepsilon q^\varepsilon = 0 & \text{in } B^\varepsilon \times (0,T), \end{cases}$$

$$\tag{2}$$

where $\dot{f} = \partial_t f$ denotes the time derivative of f and T_0 represents a reference temperature. The transmission conditions across the interfaces $S^{+,\varepsilon}$ and $S^{-,\varepsilon}$ implies the continuity of the state s^ε and of its normal dual counterpart with respect to $S^{\pm,\varepsilon}$, meaning that $[u^\varepsilon] = 0$, $[\theta^\varepsilon] = 0$, $[\sigma^\varepsilon e_3] = 0$, $[q^\varepsilon \cdot e_3] = 0$ on $S^{\pm,\varepsilon} \times (0,T)$, where $[f]$ stands for the jump function evaluated at the interface $S^{\pm,\varepsilon}$. The boundary conditions are posed on $\Gamma^\varepsilon \times (0,T)$, with $\Gamma^\varepsilon := \Gamma^{+,\varepsilon} \cup \Gamma^{-,\varepsilon}$; we recall that $\Gamma^\varepsilon = \Gamma^\varepsilon_g \cup \Gamma^\varepsilon_u$. For simplicity, we assume homogeneous boundary conditions on $\Gamma^\varepsilon_u \times (0,T)$, concerning displacements and temperature, and non-homogeneous boundary conditions on $\Gamma^\varepsilon_g \times (0,T)$, concerning surface forces $g^\varepsilon = (g^\varepsilon_i)$ and surface heat flow q^ε. Hence, one has: $\sigma^\varepsilon n^\varepsilon = g^\varepsilon$ and $-q^\varepsilon \cdot n^\varepsilon = q^\varepsilon$ on $\Gamma^\varepsilon_g \times (0,T)$, and $u^\varepsilon = 0$ and $\theta^\varepsilon = 0$ on $\Gamma^\varepsilon_u \times (0,T)$, where $n^\varepsilon = (n^\varepsilon_i)$ is the outer unit normal vector to $\partial\Omega^\varepsilon$. The initial conditions are posed in Ω^ε. Let θ^ε_{in}, u^ε_{in}, and $\dot{u}^\varepsilon_{in}$ be, respectively, the variation of temperature, the displacement, and velocity fields at time $t = 0$; one has $\theta^\varepsilon(x^\varepsilon, 0) = \theta^\varepsilon(0) = \theta^\varepsilon_{in}$, $u^\varepsilon(x^\varepsilon, 0) = u^\varepsilon(0) = u^\varepsilon_{in}$ and $\dot{u}^\varepsilon(x^\varepsilon, 0) = \dot{u}^\varepsilon(0) = \dot{u}^\varepsilon_{in}$ in Ω^ε.

Let us introduce the functional spaces $V(\Omega^\varepsilon) := \{v^\varepsilon \in H^1(\Omega^\varepsilon); \ v^\varepsilon = 0 \text{ on } \Gamma^\varepsilon_u\}$ and $\mathbf{V}(\Omega^\varepsilon) := [V(\Omega^\varepsilon)]^3$. Given a certain state $s^\varepsilon := (u^\varepsilon, \theta^\varepsilon) \in \mathbb{V}(\Omega^\varepsilon) := \mathbf{V}(\Omega^\varepsilon) \times V(\Omega^\varepsilon)$, for all test functions $r^\varepsilon = (\mathbf{v}^\varepsilon, \zeta^\varepsilon) \in \mathbb{V}(\Omega^\varepsilon)$ and for any fixed $t \in (0, T)$, we introduce the following bilinear and linear forms:

$$A^\varepsilon(s^\varepsilon, r^\varepsilon) := \int_{\Omega^\varepsilon} \left\{ \rho^\varepsilon \ddot{\mathbf{u}}^\varepsilon \cdot \mathbf{v}^\varepsilon + \sigma^\varepsilon : \mathbf{e}^\varepsilon(\mathbf{v}^\varepsilon) + \dot{S}^\varepsilon \zeta^\varepsilon - \frac{1}{T_0} \mathbf{q}^\varepsilon \cdot \nabla^\varepsilon \zeta^\varepsilon \right\} dx^\varepsilon, \tag{3}$$

$$L^\varepsilon(r^\varepsilon) := \int_{\Omega^\varepsilon_\pm} \{ \mathbf{f}^\varepsilon \cdot \mathbf{v}^\varepsilon + h^\varepsilon \zeta^\varepsilon \} dx^\varepsilon + \int_{\Gamma^\varepsilon_g} \{ \mathbf{g}^\varepsilon \cdot \mathbf{v}^\varepsilon + q^\varepsilon \zeta^\varepsilon \} d\Gamma^\varepsilon. \tag{4}$$

The variational form of the coupled thermoelastic system defined on the variable domain Ω^ε reads as follows:

$$\begin{cases} \text{Find } s^\varepsilon(t) \in \mathbb{V}(\Omega^\varepsilon), \ t \in (0, T), \ \text{such that} \\ \bar{A}^\varepsilon_-(s^\varepsilon, r^\varepsilon) + \bar{A}^\varepsilon_+(s^\varepsilon, r^\varepsilon) + \hat{A}^\varepsilon(s^\varepsilon, r^\varepsilon) = L^\varepsilon(r^\varepsilon), \end{cases} \tag{5}$$

for all $r^\varepsilon \in \mathbb{V}(\Omega^\varepsilon)$, with initial condition θ^ε_{in}, $\mathbf{u}^\varepsilon_{in}$, and $\dot{\mathbf{u}}^\varepsilon_{in}$. The coupled hyperbolic–parabolic equations associated with variation problem (5) imply a degenerate system. Hence, the standard existence theorems are not applicable. For instance, in [37,38], by applying the pseudo-monotone theory, a weak solution is provided for a sufficiently small thermal stress coupling coefficient. In [39,40], a solution to the implicit evolution equation is derived after time-differentiation of the equilibrium equation provided by sufficiently smooth data of the problem. Under suitable regularity properties of the initial data, source and boundary values, and constitutive parameters, the well-posedness of thermo-electro-elastic evolution problem is extensively discussed in [41]: the proof of existence, uniqueness, and regularity of the solution has been obtained through the Faedo–Galerkin method. The existence and uniqueness theorems have also been extended to the thermo-electro-magneto-elastic case [42] and can be easily adapted to the present coupled thermoelastic problem.

2.1. Rescaling

To study the asymptotic behavior of the solution of problem (5) when ε tends to zero, we rewrite the problem on a fixed domain Ω independent of ε. By using the approach of (author?) [17], we consider the bijection $\pi^\varepsilon : x \in \overline{\Omega} \mapsto x^\varepsilon \in \overline{\Omega}^\varepsilon$ given by

$$\pi^\varepsilon : \begin{cases} \bar{\pi}^\varepsilon(x_1, x_2, x_3) = (x_1, x_2, x_3 \mp \frac{1}{2}(1 - \varepsilon)), & \text{for all } x \in \overline{\Omega}_\pm, \\ \hat{\pi}^\varepsilon(x_1, x_2, x_3) = (x_1, x_2, \varepsilon x_3), & \text{for all } x \in \overline{B}, \end{cases} \tag{6}$$

where, after the change of variables, the adherents occupy $\Omega_\pm := \Omega^\varepsilon_\pm \pm \frac{1}{2}(1 - \varepsilon)\mathbf{e}_3$ and the interphase $B = \{x \in \mathbb{R}^3 : (x_1, x_2) \in S, |x_3| < \frac{1}{2}\}$. The sets $S_\pm = \{x \in \mathbb{R}^3 : (x_1, x_2) \in S, x_3 = \pm\frac{1}{2}\}$ denote the interfaces between B and $\Omega_\pm$ and $\Omega = \Omega_+ \cup \Omega_- \cup B$ is the rescaled configuration of the composite. Lastly, Γ_u and Γ_g indicate the images through π^ε of Γ^ε_u and Γ^ε_g (cf. Figure 1b). Consequently, $\frac{\partial}{\partial x^\varepsilon_\alpha} = \frac{\partial}{\partial x_\alpha}$ and $\frac{\partial}{\partial x^\varepsilon_3} = \frac{\partial}{\partial x_3}$ in $\Omega_\pm$, and $\frac{\partial}{\partial x^\varepsilon_\alpha} = \frac{\partial}{\partial x_\alpha}$ and $\frac{\partial}{\partial x^\varepsilon_3} = \frac{1}{\varepsilon} \frac{\partial}{\partial x_3}$ in B. In the sequel, only if necessary, $\bar{s}^\varepsilon = (\bar{\mathbf{u}}^\varepsilon, \bar{\theta}^\varepsilon)$ and $\hat{s}^\varepsilon = (\hat{\mathbf{u}}^\varepsilon, \hat{\theta}^\varepsilon)$ denote the restrictions of functions $s^\varepsilon = (\mathbf{u}^\varepsilon, \theta^\varepsilon)$ to $\Omega_\pm$ and B.

The constitutive coefficients of $\Omega^\varepsilon_\pm$ are assumed to be independent of ε, while the constitutive coefficients of B^ε present the following dependences on ε: $\hat{\lambda}^\varepsilon = \varepsilon^p \hat{\lambda}$, $\hat{\mu}^\varepsilon = \varepsilon^p \hat{\mu}$, $\hat{\beta}^\varepsilon = \varepsilon^p \hat{\beta}$, $\hat{c}^\varepsilon_v = \varepsilon^p \hat{c}_v$, and $\hat{k}^{m,\varepsilon} = \varepsilon^p \hat{k}$, with $p \in \{0, 1\}$. Two different limit behaviors are characterized according to the choice of the exponent p: by choosing $p = 1$, a model for a *soft* thermoelastic interface with low conductivity is deduced; and, when $p = 0$, a model for a *hard* thermoelastic interface with moderate conductivity is obtained. Finally, the data, unknowns, and test functions verify the following scaling assumptions: $s^\varepsilon(x^\varepsilon) = s^\varepsilon(x)$, $r^\varepsilon(x^\varepsilon) = r(x) \ x \in \Omega$, $\mathbf{f}^\varepsilon(x^\varepsilon) = \mathbf{f}(x)$, $h^\varepsilon(x^\varepsilon) = h(x) \ x \in \Omega_\pm$, $\mathbf{g}^\varepsilon(x^\varepsilon) = \mathbf{g}(x)$, $q^\varepsilon(x^\varepsilon) = q(x), x \in \Gamma_g$. Thus, $L^\varepsilon(r^\varepsilon) = L(r)$.

According to the previous hypothesis, problem (5) can be reformulated on a fixed domain Ω independent of ε. Thus, the following rescaled problem (in the sequel, we omit the explicit dependences on time t of the unknowns and data) is obtained:

$$\begin{cases} \text{Find } s^\varepsilon \in \mathbb{V}(\Omega), \ t \in (0, T), \ \text{such that} \\ \bar{A}_-(s^\varepsilon, r) + \bar{A}_+(s^\varepsilon, r) + \varepsilon^{p+1}\hat{A}(s^\varepsilon, r) = L(r), \end{cases} \tag{7}$$

for all $r \in \mathbb{V}(\Omega)$, $p \in \{0, 1\}$, with initial condition θ_{in}, $\mathbf{u}_{in}$, and $\dot{\mathbf{u}}_{in}$, where

$$\bar{A}_\pm(s^\varepsilon, r) := \int_{\Omega_\pm} \left\{ \rho^\varepsilon \ddot{\mathbf{u}}^\varepsilon \cdot \mathbf{v}^\varepsilon + \sigma^\varepsilon : \mathbf{e}(\mathbf{v}) + \dot{S}^\varepsilon \zeta - \frac{1}{T_0} \mathbf{q}^\varepsilon \cdot \nabla \zeta \right\} dx, \tag{8}$$

$$\hat{A}(s^\varepsilon, r) := \frac{1}{\varepsilon^2} a_0(s^\varepsilon, r) + \frac{1}{\varepsilon} a_1(s^\varepsilon, r) + a_2(s^\varepsilon, r), \tag{9}$$

where

$$a_0(s^\varepsilon, r) := \int_B \left\{ \hat{\mathbf{K}} \mathbf{u}^\varepsilon_{,3} \cdot \mathbf{v}_{,3} + \frac{\hat{k}}{T_0} \theta^\varepsilon_{,3} \zeta_{,3} \right\} dx, \tag{10}$$

$$a_1(s^\varepsilon, r) := \int_B \left\{ (\hat{\mathbf{K}}^\alpha)^T \mathbf{u}^\varepsilon_{,\alpha} \cdot \mathbf{v}_{,3} + \hat{\mathbf{K}}^\alpha \mathbf{u}^\varepsilon_{,3} \cdot \mathbf{v}_{,\alpha} - \hat{\beta} \theta^\varepsilon v_{3,3} + \hat{\beta} \dot{u}^\varepsilon_{3,3} \zeta \right\} dx, \tag{11}$$

$$a_2(s^\varepsilon, r) := \int_B \left\{ \hat{\mathbf{K}}^{\alpha\beta} \mathbf{u}^\varepsilon_{,\beta} \cdot \mathbf{v}_{,\alpha} - \hat{\beta}^\varepsilon \theta^\varepsilon v_{\tau,\tau} + \frac{\hat{k}}{T_0} \theta^\varepsilon_{,\alpha} \zeta_{,\alpha} + (\hat{c}_v \dot{\theta}^\varepsilon + \hat{\beta} \dot{u}^\varepsilon_{\alpha,\alpha}) \zeta \right\} dx \tag{12}$$

and

$$\hat{\mathbf{K}} := \begin{bmatrix} \hat{\mu} & 0 & 0 \\ 0 & \hat{\mu} & 0 \\ 0 & 0 & 2\hat{\mu} + \hat{\lambda} \end{bmatrix}, \ \hat{\mathbf{K}}^1 := \begin{bmatrix} 0 & 0 & \hat{\lambda} \\ 0 & 0 & 0 \\ \hat{\mu} & 0 & 0 \end{bmatrix}, \ \hat{\mathbf{K}}^2 := \begin{bmatrix} 0 & 0 & 0 \\ 0 & 0 & \hat{\lambda} \\ 0 & \hat{\mu} & 0 \end{bmatrix}, \tag{13}$$

$$\hat{\mathbf{K}}^{11} := \begin{bmatrix} 2\hat{\mu} + \hat{\lambda} & 0 & 0 \\ 0 & \hat{\mu} & 0 \\ 0 & 0 & \hat{\mu} \end{bmatrix}, \ \hat{\mathbf{K}}^{22} := \begin{bmatrix} \hat{\mu} & 0 & 0 \\ 0 & 2\hat{\mu} + \hat{\lambda} & 0 \\ 0 & 0 & \hat{\mu} \end{bmatrix}, \tag{14}$$

$$\hat{\mathbf{K}}^{12} := \begin{bmatrix} 0 & \hat{\lambda} & 0 \\ \hat{\mu} & 0 & 0 \\ 0 & 0 & 0 \end{bmatrix}, \ \hat{\mathbf{K}}^{21} = (\hat{\mathbf{K}}^{12})^T. \tag{15}$$

Now, an asymptotic analysis of the rescaled problem (7) can be performed. Since the rescaled problem (7) has a polynomial structure with respect to the small parameter ε, we can look for the solution s^ε of the problem as a series of powers of ε:

$$s^\varepsilon = s^0 + \varepsilon s^1 + \varepsilon^2 s^2 + \ldots, \ \bar{s}^\varepsilon = \bar{s}^0 + \varepsilon \bar{s}^1 + \varepsilon^2 \bar{s}^2 + \ldots, \ \hat{s}^\varepsilon = \hat{s}^0 + \varepsilon \hat{s}^1 + \varepsilon^2 \hat{s}^2 + \ldots . \tag{16}$$

where $\bar{s}^\varepsilon = s^\varepsilon \circ \bar{\pi}^\varepsilon$ and $\hat{s}^\varepsilon = s^\varepsilon \circ \hat{\pi}^\varepsilon$. By substituting (16) into the rescaled problem (7), and by identifying the terms with identical power of ε, as customary, a set of variational problems is obtained to be solved in order to characterize the limit thermoelastic state s^0, the first-order corrector term s^1 and their associated limit problem, for $p \in \{0, 1\}$. The order 1 can be considered as a corrector term of the order 0, giving a better approximation of the initial model.

3. The Soft Thermoelastic Interface Model

In this section, the limit model for a soft thermoelastic interface model, corresponding to an adhesive which is weaker with respect to the adherents, is derived. By choosing $p = 1$ and injecting (16) into (7), the following set of variational problems $\mathcal{P}_q$ is obtained:

$$\begin{cases} \mathcal{P}_0: & \bar{A}_-(s^0, r) + \bar{A}_+(s^0, r) + a_0(s^0, r) = L(r), \\ \mathcal{P}_1: & \bar{A}_-(s^1, r) + \bar{A}_+(s^1, r) + a_0(s^1, r) + a_1(s^0, r) = 0, \\ \mathcal{P}_q: & \bar{A}_-(s^q, r) + \bar{A}_+(s^q, r) + a_0(s^q, r) + a_1(s^{q-1}, r) + a_2(s^{q-2}, r) = 0, \ q \geqslant 2 \end{cases} \tag{17}$$

In the sequel, the limit problems at order 0 and order 1 are presented, by skipping all the mathematical technicalities involved in the solution of problems $\mathcal{P}_q$ (see [1] for a detailed description of the asymptotic analysis).

- Order 0 model

Governing equations   **Transmission conditions on** $S_{\pm}$

$$
\begin{cases}
\rho\ddot{\bar{\mathbf{u}}}^0 - \mathrm{div}\,\bar{\sigma}^0 = \mathbf{f} & \text{in } \Omega_{\pm}, \\
\dot{\check{S}}^0 + \frac{1}{T_0}\mathrm{div}\,\bar{\mathbf{q}}^0 = h & \text{in } \Omega_{\pm}, \\
\bar{\sigma}^0\mathbf{n} = \mathbf{g} & \text{on } \Gamma_g \\
-\bar{\mathbf{q}}^0\cdot\mathbf{n} = q & \text{on } \Gamma_g, \\
s^0 = 0 & \text{on } \Gamma_u,
\end{cases}
\qquad
\begin{cases}
[\bar{\mathbf{u}}^0] = \hat{\mathbf{K}}^{-1}\langle\bar{\sigma}^0\mathbf{e}_3\rangle, \\
[\bar{\theta}^0] = -\frac{T_0}{\hat{k}}\langle\bar{\mathbf{q}}^0\cdot\mathbf{e}_3\rangle, \\
[\bar{\sigma}^0\mathbf{e}_3] = 0, \\
[\bar{\mathbf{q}}^0\cdot\mathbf{e}_3] = 0.
\end{cases}
\tag{18}
$$

- Order 1 model

Governing equations   **Transmission conditions on** $S_{\pm}$

$$
\begin{cases}
\rho\ddot{\bar{\mathbf{u}}}^1 - \mathrm{div}\,\bar{\sigma}^1 = 0 & \text{in } \Omega_{\pm}, \\
\dot{\check{S}}^1 + \frac{1}{T_0}\mathrm{div}\,\bar{\mathbf{q}}^1 = 0 & \text{in } \Omega_{\pm}, \\
\bar{\sigma}^1\mathbf{n} = 0 & \text{on } \Gamma_g \\
\bar{\mathbf{q}}^1\cdot\mathbf{n} = 0 & \text{on } \Gamma_g, \\
s^1 = 0 & \text{on } \Gamma_u,
\end{cases}
\qquad
\begin{cases}
[\bar{\mathbf{u}}^1] = \hat{\mathbf{K}}^{-1}\{\langle\bar{\sigma}^1\mathbf{e}_3\rangle - (\hat{\mathbf{K}}^\alpha)^T\langle\bar{\mathbf{u}}^0\rangle_{,\alpha} + \hat{\beta}\langle\bar{\theta}^0\rangle\mathbf{e}_3\}, \\
[\bar{\theta}^1] = -\frac{T_0}{\hat{k}}\langle\bar{\mathbf{q}}^1\cdot\mathbf{e}_3\rangle, \\
[\bar{\sigma}^1\mathbf{e}_3] = -\mathbf{K}^\alpha[\bar{\mathbf{u}}^0]_{,\alpha}, \\
[\bar{\mathbf{q}}^1\cdot\mathbf{e}_3] = -\hat{\beta}[\dot{\bar{u}}^0_3],
\end{cases}
$$

$$\tag{19}$$

where $\langle f\rangle := \frac{1}{2}(f(\tilde{x},+(1/2)^+) + f(\tilde{x},-(1/2)^-)$ and $[f] := f(\tilde{x},+(1/2)^+) - f(\tilde{x},-(1/2)^-)$, $\tilde{x} := (x_\alpha) \in S$ denote, respectively, the mean value and the jump functions at the interfaces. The soft thermoelastic interface models at order 0 and order 1 present various similarities, compared with the linear elastic case [29]. At order 0, from a mechanical point of view, the interface behaves as linear springs reacting to the jump between the top and bottom displacements and temperature, while the traction vector and normal heat flow are remains continuous. The order 1 model provides a mixed contact law, expressed by a concurrent discontinuity in terms of thermoelastic state, traction vector, and normal heat flow. The order 1 transmission conditions can be also rewritten in terms of $\langle\bar{\mathbf{q}}^1\cdot\mathbf{e}_3\rangle$ and $\langle\bar{\sigma}^1\mathbf{e}_3\rangle$, as follows:

$$
\begin{cases}
\langle\bar{\sigma}^1\mathbf{e}_3\rangle = \hat{\mathbf{K}}[\bar{\mathbf{u}}^1] + (\hat{\mathbf{K}}^\alpha)^T\langle\bar{\mathbf{u}}^0\rangle_{,\alpha} - \hat{\beta}\langle\bar{\theta}^0\rangle\mathbf{e}_3, \\
\langle\bar{\mathbf{q}}^1\cdot\mathbf{e}_3\rangle = -\frac{\hat{k}}{T_0}[\bar{\theta}^1].
\end{cases}
\tag{20}
$$

The jump and mean values of the traction vector and normal heat flow at the interface depend on s^0 and are analogous to those obtained for the soft elastic case in [28]. It is interesting to notice that, at order 1, the jump of the heat flow at the interface inside the intermediate layer depend on the variation in time of the normal displacement u_3.

4. The Hard Thermoelastic Interface Model

In this section, the limit model for a hard thermoelastic interface, corresponding to an intermediate layer having the same rigidities of the top and bottom bodies, is derived. Let $p = 0$, the asymptotic expansion (16) is inserted in (7), and the following set of variational problems $\mathcal{P}_q$ is obtained:

$$
\begin{cases}
\mathcal{P}_{-1}: & a_0(s^0,r) = 0, \\
\mathcal{P}_0: & \bar{A}_-(s^0,r) + \bar{A}_+(s^0,r) + a_0(s^1,r) + a_1(s^0,r) = L(r), \\
\mathcal{P}_1: & \bar{A}_-(s^1,r) + \bar{A}_+(s^1,r) + a_0(s^2,r) + a_1(s^1,r) + a_2(s^0,r) = 0, \\
\mathcal{P}_q: & \bar{A}_-(s^q,r) + \bar{A}_+(s^q,r) + a_0(s^{q+1},r) + a_1(s^q,r) + a_2(s^{q-1},r) = 0, \; q \geqslant 2
\end{cases}
\tag{21}
$$

A detailed equivalent analysis on the solution of the variational problems $\mathcal{P}_q$ can be found in [1]. In the sequel, the limit problems at order 0 and order 1 are presented.

- Order 0 model

Governing equations   **Transmission conditions on** S_+

$$
\begin{cases}
\rho \ddot{\bar{\mathbf{u}}}^0 - \operatorname{div} \bar{\sigma}^0 = \mathbf{f} & \text{in } \Omega_\pm, \\
\dot{\bar{S}}^0 + \frac{1}{T_0} \operatorname{div} \bar{\mathbf{q}}^0 = h & \text{in } \Omega_\pm, \\
\bar{\sigma}^0 \mathbf{n} = \mathbf{g} & \text{on } \Gamma_g, \\
-\bar{\mathbf{q}}^0 \cdot \mathbf{n} = q & \text{on } \Gamma_g, \\
s^0 = 0 & \text{on } \Gamma_u,
\end{cases}
\qquad
\begin{cases}
[\bar{\mathbf{u}}^0] = 0, \\
[\bar{\theta}^0] = 0, \\
[\bar{\sigma}^0 \mathbf{e}_3] = 0, \\
[\bar{\mathbf{q}}^0 \cdot \mathbf{e}_3] = 0.
\end{cases}
\tag{22}
$$

- Order 1 model

Governing equations   **Transmission conditions on** S_+

$$
\begin{cases}
\rho \ddot{\bar{\mathbf{u}}}^1 - \operatorname{div} \bar{\sigma}^1 = 0 & \text{in } \Omega_\pm, \\
\dot{\bar{S}}^1 + \frac{1}{T_0} \operatorname{div} \bar{\mathbf{q}}^1 = 0 & \text{in } \Omega_\pm, \\
\bar{\sigma}^1 \mathbf{n} = 0 & \text{on } \Gamma_g, \\
\bar{\mathbf{q}}^1 \cdot \mathbf{n} = 0 & \text{on } \Gamma_g, \\
s^1 = 0 & \text{on } \Gamma_u,
\end{cases}
$$

$$
\begin{cases}
[\bar{\mathbf{u}}^1] = \hat{\mathbf{K}}^{-1}\{\langle \bar{\sigma}^0 \mathbf{e}_3 \rangle - (\hat{\mathbf{K}}^\alpha)^T \bar{\mathbf{u}}^0_{,\alpha} + \hat{\beta}\bar{\theta}^0 \mathbf{e}_3\}, \\
[\bar{\theta}^1] = -\frac{T_0}{\hat{k}} \langle \bar{\mathbf{q}}^0 \cdot \mathbf{e}_3 \rangle, \\
[\bar{\sigma}^1 \mathbf{e}_3] = -\left(\hat{\mathbf{K}}^\alpha \hat{\mathbf{K}}^{-1} \langle \bar{\sigma}^0 \mathbf{e}_3 \rangle_{,\alpha} + \hat{\mathbf{L}}^{\alpha\beta} \bar{\mathbf{u}}^0_{,\alpha\beta} + \right. \\
\qquad\qquad \left. + \hat{\beta}(\hat{\mathbf{K}}^\alpha \hat{\mathbf{K}}^{-1} \bar{\theta}^0_{,\alpha} \mathbf{e}_3 - \bar{\theta}^0_{,\tau} \mathbf{e}_\tau) \right), \\
[\bar{\mathbf{q}}^1 \cdot \mathbf{e}_3] = -\left(\frac{\hat{\beta}}{\hat{\lambda}+2\hat{\mu}} \langle \dot{\bar{\sigma}}^0_{33} \rangle + \dot{\bar{\Sigma}}^0 - \frac{\hat{k}}{T_0} \Delta_s \bar{\theta}^0 \right),
\end{cases}
\tag{23}
$$

where $\hat{\mathbf{L}}^{\alpha\beta} := \hat{\mathbf{K}}^{\alpha\beta} - \hat{\mathbf{K}}^\beta \hat{\mathbf{K}}^{-1}(\hat{\mathbf{K}}^\alpha)^T$, $\tilde{\Sigma}^0 := \tilde{\beta}\bar{u}^0_{\alpha,\alpha} + \tilde{c}_v \bar{\theta}^0$, with $\tilde{\beta} := \frac{2\hat{\mu}\hat{\beta}}{\hat{\lambda}+2\hat{\mu}}$ and $\tilde{c}_v := \hat{c}_v + \frac{\hat{\beta}^2}{\hat{\lambda}+2\hat{\mu}}$, Δ_s denotes the two-dimensional Laplacian operator. Note that, in this case, $\langle \bar{\theta}^0 \rangle = \bar{\theta}^0$ and $\langle \bar{\mathbf{u}}^0 \rangle = \bar{\mathbf{u}}^0$.

The hard thermoelastic interface problems above present the same structures of the analogous linear elastic hard interface models [26–28]. Concerning the order 0, the transmission conditions provide a continuity of the thermoelastic state and of its conjugated counterpart, which is typical for adhesives having the same rigidity properties of the adherents. In this case, the upper and lower bodies are perfectly bonded together. At order 1, a mixed interface model is obtained, characterized by a jump of the state and traction vector depending on the values of the thermoelastic state and traction vector at order 0. These order 0 terms are known since they have been determined in the previous problem and they appear in the formulation as source terms. The interface conditions at order 1 can be interpreted as the two-dimensional coupled thermoelastic problem defined on the plane of the interface.

5. Generalized Interface Transmission Conditions

In [1,28], it has been shown that it is possible to obtain a condensed form of transmission conditions summarizing both the orders 0 and 1 of the soft and hard cases in only one couple of equations in terms of the jump of the displacement field and tractions at the interface. Equivalently, it is possible to define an implicit general thermoelastic interface law starting from the hard case, comprising the order 0 and order 1 soft and hard thermoelastic interface models.

To this end, by denoting by $\tilde{s}^\varepsilon := \bar{s}^0 + \varepsilon\bar{s}^1$, $\tilde{\sigma}^\varepsilon := \bar{\sigma}^0 + \varepsilon\bar{\sigma}^1$ and $\tilde{\mathbf{q}}^\varepsilon := \bar{\mathbf{q}}^0 + \varepsilon\bar{\mathbf{q}}^1$, suitable approximations of s^ε, $\bar{\sigma}^\varepsilon$ and $\bar{\mathbf{q}}^\varepsilon$, respectively, and following the approach developed in [1,28], one can obtain the implicit form of the transmission conditions:

$$
\begin{cases}
\langle \tilde{\sigma}^\varepsilon \mathbf{e}_3 \rangle = \frac{1}{\varepsilon} \hat{\mathbf{K}}[\tilde{\mathbf{u}}^\varepsilon] + (\hat{\mathbf{K}}^\alpha)^T \langle \tilde{\mathbf{u}}^\varepsilon \rangle_{,\alpha} - \hat{\beta}\langle \tilde{\theta}^\varepsilon \rangle \mathbf{e}_3, \\
\langle \tilde{\mathbf{q}}^\varepsilon \cdot \mathbf{e}_3 \rangle = -\frac{\hat{k}}{\varepsilon T_0}[\tilde{\theta}^\varepsilon], \\
[\tilde{\sigma}^\varepsilon \mathbf{e}_3] = -\hat{\mathbf{K}}^\alpha [\tilde{\mathbf{u}}^\varepsilon]_{,\alpha} - \varepsilon \hat{\mathbf{K}}^{\alpha\beta} \langle \tilde{\mathbf{u}}^\varepsilon \rangle_{,\alpha\beta} + \varepsilon\hat{\beta}\langle \tilde{\theta}^\varepsilon \rangle_{,\alpha} \mathbf{e}_\alpha, \\
[\tilde{\mathbf{q}}^\varepsilon \cdot \mathbf{e}_3] = -\left(\hat{\beta}[\dot{\tilde{u}}^\varepsilon_3] + \varepsilon\langle \dot{\tilde{\Sigma}}^\varepsilon \rangle - \varepsilon\frac{\hat{k}}{T_0} \Delta_s \langle \tilde{\theta}^\varepsilon \rangle \right),
\end{cases}
\tag{24}
$$

91

with $\langle \Sigma^\varepsilon \rangle := \hat{c}_v \langle \tilde{\theta}^\varepsilon \rangle + \hat{\beta} \langle \tilde{u}_\alpha^\varepsilon \rangle_{,\alpha}$.

To write the variational formulation of the general coupled thermoelastic interface problem, the expression of the general transmission conditions presented in (24) is employed. In what follows, for the sake of simplicity, the indices ε and symbol $(\tilde{\cdot})$ are omitted. Let us write the variational form of the equilibrium equations on each sub-domain Ω_+ and Ω_-. The sum of the two equations leads to

$$\int_{\Omega_\pm} \left\{ \rho \ddot{\mathbf{u}} \cdot \mathbf{v} + \sigma : \mathbf{e}(\mathbf{v}) + \dot{\mathcal{S}}\xi - \frac{1}{T_0}\mathbf{q} \cdot \nabla\xi \right\} dx -$$
$$- \int_S \left\{ \sigma(\tilde{x}, 0^+)\mathbf{n}(\tilde{x}, 0^+) \cdot \mathbf{v} + \sigma(\tilde{x}, 0^-)\mathbf{n}(\tilde{x}, 0^-) \cdot \mathbf{v} \right\} d\Gamma + \tag{25}$$
$$+ \int_S \left\{ \mathbf{q}(\tilde{x}, 0^+) \cdot \mathbf{n}(\tilde{x}, 0^+)\xi + \mathbf{q}(\tilde{x}, 0^-) \cdot \mathbf{n}(\tilde{x}, 0^-)\xi \right\} d\Gamma = L(r),$$

which can be written

$$\int_{\Omega_\pm} \left\{ \rho \ddot{\mathbf{u}} \cdot \mathbf{v} + \sigma : \mathbf{e}(\mathbf{v}) + \dot{\mathcal{S}}\xi - \frac{1}{T_0}\mathbf{q} \cdot \nabla\xi \right\} dx + \int_S [\sigma \mathbf{e}_3 \cdot \mathbf{v}] - [\mathbf{q} \cdot \mathbf{e}_3 \xi] d\tilde{x} = L(r), \tag{26}$$

letting $\mathbf{e}_3 = \mathbf{n}(\tilde{x}, 0^-) = -\mathbf{n}(\tilde{x}, 0^+)$ and $d\Gamma = d\tilde{x}$. Then, using the property $[ab] = \langle a \rangle [b] + [a]\langle b \rangle$ and relations (24), and after an integration by parts, one has

$$\begin{cases} \text{Find } s \in \mathbb{W}(\tilde{\Omega}), \ \tilde{\Omega} := \Omega_+ \cup S \cup \Omega_-, \ t \in (0, T), \text{ such that} \\ \bar{A}_-(s, r) + \bar{A}_+(s, r) + \mathcal{A}(s, r) = \mathcal{L}(r), \end{cases} \tag{27}$$

for all $r \in \mathbb{W}(\tilde{\Omega})$, where $\mathbb{W}(\tilde{\Omega}) := \mathbf{W}(\tilde{\Omega}) \times W(\tilde{\Omega})$, with $W(\tilde{\Omega}) := \{r \in H^1(\tilde{\Omega}), \ r|_S \in H^1(S), \ r = 0 \text{ on } \Gamma_u\}$, $\mathbf{W}(\tilde{\Omega}) := [W(\tilde{\Omega})]^3$, and

$$\mathcal{A}(s, r) := \int_S \left\{ \frac{1}{\varepsilon}\hat{\mathbf{K}}[\mathbf{u}] \cdot [\mathbf{v}] + (\hat{\mathbf{K}}^\alpha)^T \langle \mathbf{u} \rangle_{,\alpha} \cdot [\mathbf{v}] + \hat{\mathbf{K}}^\alpha[\mathbf{u}] \cdot \langle \mathbf{v} \rangle_{,\alpha} + \varepsilon \hat{\mathbf{K}}^{\alpha\beta} \langle \mathbf{u} \rangle_{,\beta} \cdot \langle \mathbf{v} \rangle_{,\alpha} - \right.$$
$$\left. -\hat{\beta}\langle \theta \rangle [v_3] - \varepsilon \hat{\beta} \langle \theta \rangle \langle v_\alpha \rangle_{,\alpha} + \frac{1}{\varepsilon}\frac{\hat{k}}{T_0}[\theta][\xi] + \varepsilon \frac{\hat{k}}{T_0}\langle \theta \rangle_{,\alpha} \langle \xi \rangle_{,\alpha} + \hat{\beta}[\dot{u}_3]\langle \xi \rangle + \varepsilon \langle \dot{\Sigma} \rangle \langle \xi \rangle \right\} d\tilde{x}, \tag{28}$$

$$\mathcal{L}(r) := \int_{\Omega_\pm} \{ \mathbf{f} \cdot \mathbf{v} + h\xi \} dx + \int_{\Gamma_g} \{ \mathbf{g} \cdot \mathbf{v} + q\xi \} d\Gamma + \int_{\partial S} \{ \mathbf{F} \cdot \langle \mathbf{v} \rangle + \mathcal{H}\langle \xi \rangle \} d\gamma, \tag{29}$$

where $\langle \Sigma \rangle := \hat{c}_v \langle \theta \rangle + \hat{\beta} \langle u_\alpha \rangle_{,\alpha}$, $(\hat{\mathbf{K}}^\alpha[\mathbf{u}] + \varepsilon \hat{\mathbf{K}}^{\alpha\beta}\langle \mathbf{u} \rangle_{,\beta} - \varepsilon \hat{\beta}\langle \theta \rangle \mathbf{e}_\alpha)\nu_\alpha := \mathbf{F}$ and $\varepsilon \frac{\hat{k}}{T_0}\langle \theta \rangle_{,\alpha}\nu_\alpha := \mathcal{H}$ denote the loads on the lateral boundary of the interface ∂S, with outer unit normal vector (ν_α) (see [1]).

6. FEM Implementation

The numerical simulations were carried out by means of the finite element method, discretizing the variational problem (27). This helped validate the proposed asymptotic approach. The FEM analysis was performed considering the coupled dynamic thermoelastic problem and comparing the solution of the three-phase model (two adherents and adhesive) with the generalized interface (two adherents + interface). The problem was solved employing the software GetFem++ (see [35,36] for more details), with a standard linear solver (conjugate gradient). For that purpose, standard piecewise linear finite elements were considered.

Let us consider a thermoelastic laminated plate occupying a 3D domain defined by $\Omega^\varepsilon = [0, L_1] \times [0, L_2] \times [0, 2h + \varepsilon]$, with $h = 1.\text{cm}$, $L_1/h = 10$, $L_2/h = 5$. (see Figure 2). Clearly, with self-explanatory notation, $x_1 = x$, $x_2 = y$, and $x_3 = z$. The adherents are made of Material 1, while the adhesive is constituted by Material 2.

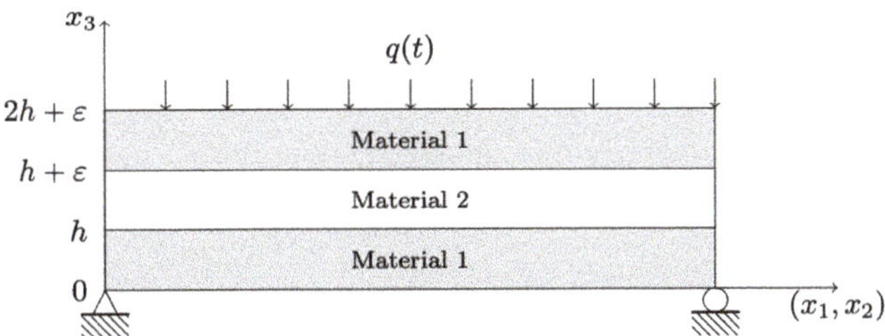

Figure 2. The 3D geometry of the thermoelastic laminated plate represented in the plane (x_1, x_3).

Simply supported boundary conditions are considered on the bottom edges of the composite plate. The plate is subject to a thermal shock $\mathbf{q} \cdot \mathbf{n}|_{\Gamma_{top}} = q(t) = ate^{-bt}$ on the top face, with $a = 30$ and $b = 0.8$, whose plot is illustrated in Figure 3. The bottom face is thermally insulated $\mathbf{q} \cdot \mathbf{n}|_{\Gamma_{bottom}} = 0$. No volume or surface mechanical loads were applied $\mathbf{f} = \mathbf{g} = 0$.

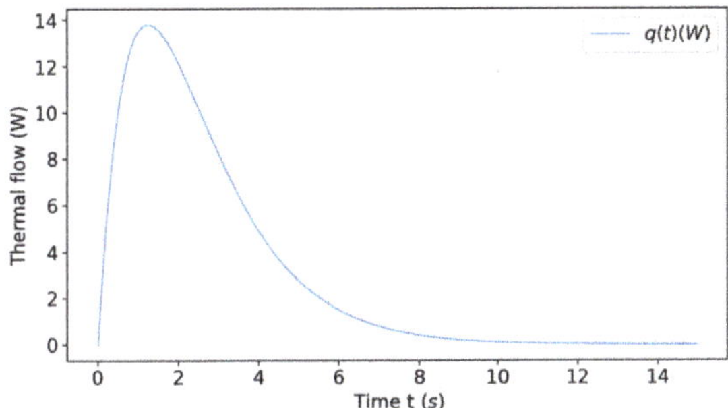

Figure 3. Applied heat flow shock .

The FEM discretization was carried out using piecewise linear finite elements on hexahedrons, with 7280 nodes (29,203 degrees of freedom) for the three-phase problem and 5824 nodes (23,635 degrees of freedom) for the problem with the generalized interface law. The time discretization was realized using a Newmark-beta scheme with $\beta = 0.25$ and $\gamma = 0.5$ for the second time derivative of the displacements, and a Crank–Nicholson scheme for the first derivative of the thermodynamic entropy. It is worth noting that both methods are unconditionally stable and of order two in time. The time step of discretization is equal to $\delta t = 0.1$ s.

The numerical example considers a composite plate, in which the adherents and the adhesive have very different thermo-mechanical properties. Material 1 is aluminum (Al), while Material 2 is a polyvinyl chloride (PVC) foam. The constitutive parameters are listed in Table 1.

Table 1. Thermoelastic material properties for Al and PVC Foam.

Material 1: Al			Material 2: PVC Foam		
ρ_1	2700	[kg/m^3]	ρ_2	250	[kg/m^3]
E_1	72.4	[GPa]	E_2	0.28	[GPa]
ν_1	0.32		ν_2	0.40	
α_1	40.0	[μm/m K]	α_2	22.4	[μm/m K]
k_1	122.2	[W/mK]	k_2	0.05	[W/mK]
c_v^1	900	[J/kg K]	c_v^2	1900	[J/kg K]

To evaluate the accuracy of the asymptotic analysis, the influence of the relative thickness ε/h, for fixed time instants, on the L^2-relative error was investigated. The L^2-relative errors $\frac{\|\mathbf{u}^\varepsilon - \mathbf{u}\|_{L^2}}{\|\mathbf{u}\|_{L^2}}$ and $\frac{\|\theta^\varepsilon - \theta\|_{L^2}}{\|\theta\|_{L^2}}$ was computed taking into account the solution $(\mathbf{u}^\varepsilon, \theta^\varepsilon)$ of the initial three-phase problem, discretized with a FE mesh, and the solution $(\mathbf{u}, \theta)$ of the interface problem (27). Tables 2 and 3 report the relative error values for increasing time and vanishing relative thickness.

Table 2. Relative error $\frac{\|\mathbf{u}^\varepsilon - \mathbf{u}\|_{L^2}}{\|\mathbf{u}\|_{L^2}}$.

$\frac{\varepsilon}{h}$ / t	5	10	15	20
0.1	4.20×10^{-03}	1.89×10^{-03}	1.58×10^{-03}	1.60×10^{-03}
0.05	7.55×10^{-04}	3.44×10^{-04}	2.76×10^{-04}	3.45×10^{-04}
0.01	2.25×10^{-05}	3.48×10^{-06}	4.52×10^{-06}	4.56×10^{-06}

Table 3. Relative error $\frac{\|\theta^\varepsilon - \theta\|_{L^2}}{\|\theta\|_{L^2}}$.

$\frac{\varepsilon}{h}$ / t	5	10	15	20
0.1	1.56×10^{-03}	5.16×10^{-04}	7.57×10^{-04}	4.97×10^{-04}
0.05	1.07×10^{-04}	2.83×10^{-04}	9.94×10^{-05}	2.27×10^{-05}
0.01	9.88×10^{-08}	5.58×10^{-07}	1.78×10^{-08}	7.43×10^{-10}

The convergence diagrams of the the relative L^2-norms of the displacements and temperatures, obtained with the three-phase problem and the reduced interface problem, are plotted in Figure 4, as the thickness ratio ε/h tends to zero, at time $t = 1$ s and $t = 10$ s, respectively. Moreover, the evolution in time of the L^2-relative is reported in Figure 5.

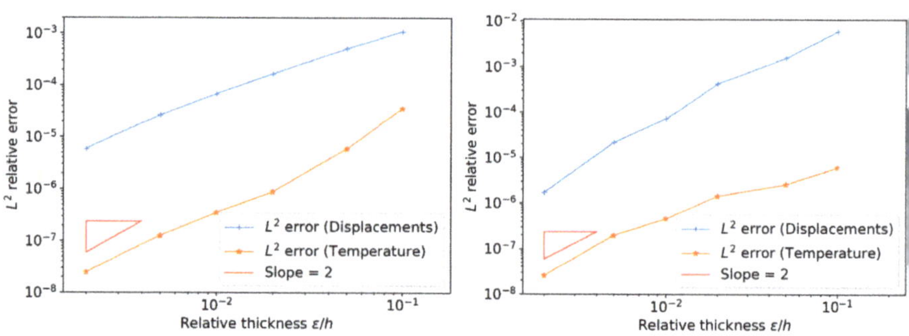

Figure 4. Convergence diagrams with respect to ε/h for: $t = 1$ s (**left**); and $t = 10$ s (**right**).

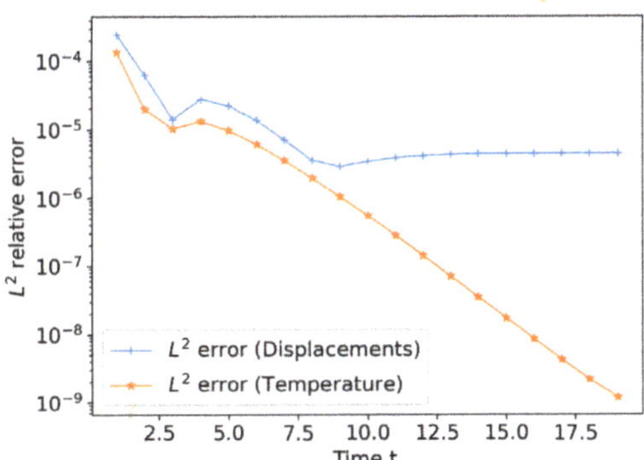

Figure 5. Evolution of the relative error with respect to the time ($\varepsilon/h = 0.01$).

From the results in Tables 2 and 3 and, especially, Figure 4, it can be noticed that, by decreasing the thickness ratio ε/h, the relative errors present an immediate reduction for fixed times. The convergence rate is of the order $(\varepsilon/h)^2$ and remains constant for increasing time instants. As illustrated in Figure 5, the evolution in time of the L^2-relative error, for fixed ε/h, becomes approximately steady after $t = 10$ s for the displacement field, while it presents a decreasing trend concerning the temperature field. Besides, even for a relative thickness $\varepsilon/h = 0.1$, at time $t = 10$ s, the relative error is close to about 1.89×10^{-3}, for the displacement field, and about 5.16×10^{-4}, for the temperature field. Hence, the proposed general thermoelastic interface model provides an acceptable solution and it is able to correctly approximate the solution $(\mathbf{u}^\varepsilon, \theta^\varepsilon)$ of the three-phase problem. Moreover, the reduced model can also be employed for moderately thick adhesives.

In the sequel, the numerical results obtained by solving the general interface model are presented, considering a relative thickness of $\varepsilon/h = 0.01$. Following the approach by [43], hereinafter, the results are provided using dimensionless variables:

- $\mathbf{U}(X_1, X_2, X_3, t) := \frac{1-\nu_1}{\ell(1+\nu_1)\alpha_1 T_0} \mathbf{u}(x_1, x_2, x_3, t),$
- $\Theta(X_1, X_2, X_3, t) := \frac{\theta(x_1, x_2, x_3, t) - T_0}{T_0},$
- $\Sigma_{ij}(X_1, X_2, X_3, t) := \frac{1}{\rho_1 V} \sigma_{ij}(x_1, x_2, x_3, t)$

where $X_i = x_i/\ell$, $t = \frac{V}{\ell}$, and ℓ and V are defined by

$$V = \sqrt{\frac{E_1(1-\nu_1)}{(1+\nu_1)(1-2\nu_1)\rho_1}}, \quad \ell = \frac{k_1}{\rho_1 c_v^1 V}. \tag{30}$$

Let us notice that the domain Ω^ε is chosen such that $X_1 \in [0; 10]$, $X_2 \in [0; 10]$ and $X_3 \in [0; 1]$.

Figure 6 represents the trend of the displacement U_3 and temperature Θ, evaluated along X_3 on the orthogonal fiber to the mid-plane of the interface at point $(\overline{X}_1 = 6, \overline{X}_2 = 6)$, for given times. The plot shows that, after the thermal shock, the displacement U_3 evolves in opposite directions within the adherents: the composite laminated plate tends to expand and contract itself along the through-the-thickness axis. On the other hand, the temperature field Θ remains constant along the X_3-axis within the adherents, for given times, reaching a steady value after a certain time interval. As expected, the plots also report a jump of the state fields (U_3, Θ) in correspondence of the intermediate layer, and, thus, the adhesive behaves as a soft thermoelastic interface. This is mainly due to the material properties of the adhesive, which are smaller with respect to the those of the adherents.

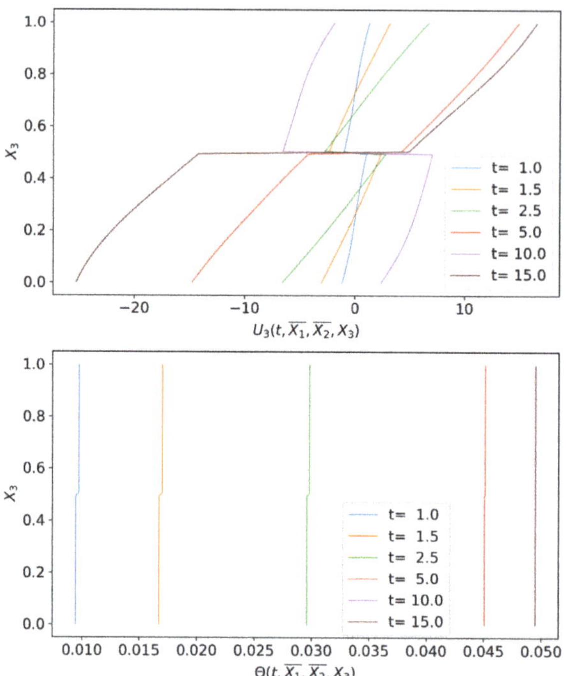

Figure 6. Displacement U_3 and temperature Θ along the X_3-axis, on a fiber $(\overline{X}_1 = 6, \overline{X}_2 = 6, X_3)$, for given times .

Figure 7 illustrates the evolution of the displacement field **U** and temperature Θ with respect to the time t, at a given point $\overline{X} = (6.5, 3.5, 0)$, placed on the bottom face of the composite plate. As expected, the thermal shock induces an oscillatory trend concerning the displacements. Conversely, the temperature evolves to a steady state, corresponding to a constant value, after a sudden increase related to the thermal shock application.

Figure 8 represents the trend of the stresses Σ_{33} and Σ_{13}, evaluated along X_3 on the orthogonal fiber to the mid-plane of the interface at point $(\overline{X}_1 = 6, \overline{X}_2 = 6)$, for given times. The plot shows that, after the thermal shock, the stress Σ_{33} remains constant along the X_3-axis within the adherents. In this particular case, the thermal contribution to Σ_{33} is predominant with respect to the elastic one, i.e., $\Sigma_{33} \approx -B(\Theta + 1)$: indeed, their diagrams present analogous trends and differ for a constant of proportionality $B := \frac{\beta T_0}{\rho_1 V}$ (see Figures 6 and 8). The stress Σ_{13} presents an oscillating behavior along X_3 inside the adherents, but its contribution is negligible compared with Σ_{33}. Moreover, the normal Σ_{33} and shear stresses Σ_{13}, evaluated at the top and bottom faces of the intermediate layer, are very similar and, thus, their jump almost vanishes. This is typical of soft interface models, in which the thermoelastic state presents a discontinuity, while its conjugated quantities (traction vector and normal heat flow) are continuous across the interface

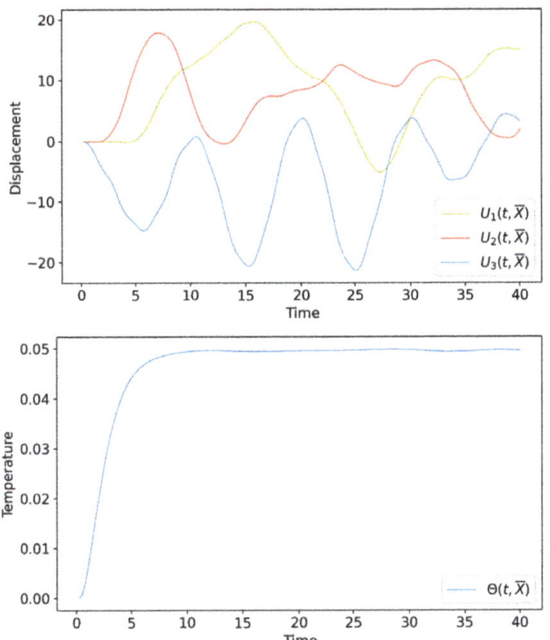

Figure 7. Displacement field $\mathbf{U} = (U_i)$ and temperature Θ versus time t, at a given point $\overline{X} = (6.5, 3.5, 0)$.

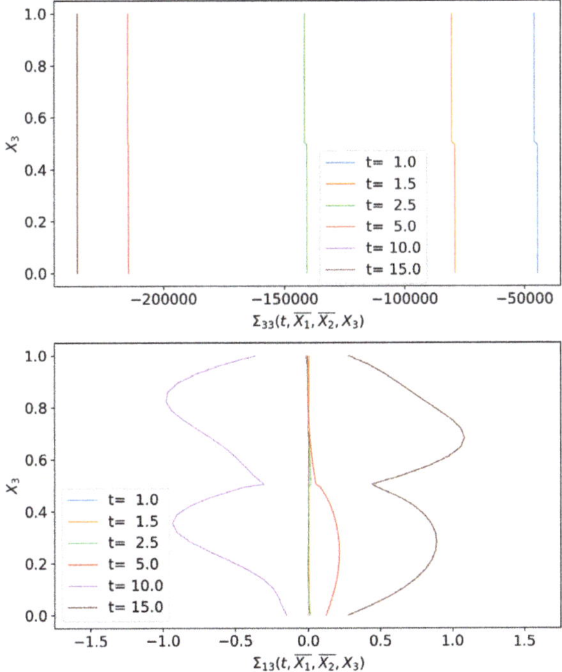

Figure 8. Stresses Σ_{33} and Σ_{13} along the X_3-axis, on a fiber $(\overline{X}_1 = 6, \overline{X}_2 = 6, X_3)$, for given times.

Figure 9 shows a comparison between the evolution in time of $U_3(\overline{X}, t)$, at a given point $\overline{X} = (6.5, 3.5, 0.)$, of a homogeneous three-layer plate, made of aluminum, and the Al/PVC composite plate. Although the thickness ratio is small ($\varepsilon/h = 0.01$), the effect of the adhesive becomes relevant concerning the response of the plate to the thermal shock. The homogeneous plate appears to be stiffer with respect to the composite one, which manifests a significant amplitude and period increase of the U_3 motion.

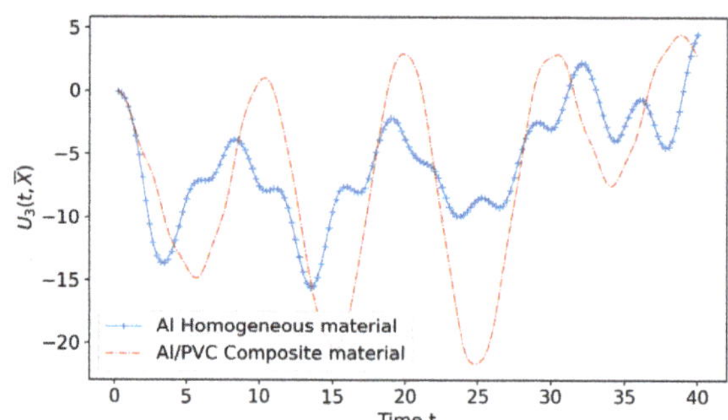

Figure 9. Evolution in time of $U_3(\overline{X}, t)$ at a given point $\overline{X} = (6.5, 3.5, 0.)$ for a homogeneous Al-plate and an Al/PVC composite plate.

7. Concluding Remarks

General imperfect interface conditions are proposed in the framework of coupled thermoelasticity, simulating the thermomechanical behavior of a thin-bonded joint. The approach used to obtain the transmission conditions is based on the asymptotic expansions method. Zero- and higher-order interface models are derived for soft and hard interphases. Following [1], a general transmission law, comprising the two regimes (soft and hard) at the various order, is derived. To assess the validity of the previous asymptotic approach, numerical simulations were developed using a finite element method, which generalizes an analogous methodology to dynamical coupled thermoelasticity, already proposed in [2] in the framework of piezoelectricity. The numerical example consisted of a thermoelastic composite three-layer aluminum plate, with a PVC adhesive, subject to a thermal shock. Two different configurations were considered: the first one consisted of an initial three-phase problem, while the second one took into account the FE discretized form of interface problem (27). The most significant fields (displacement and temperature) and their L^2-relative errors were then computed and compared to test the validity of the proposed interface laws and the accuracy of the asymptotic model. The proposed general thermoelastic interface model provides an acceptable solution and it is able to correctly approximate the solution of the three-phase problem. These findings clearly indicate that the approach of substituting the interphase with the proposed interface law provides a robust modeling for the composite.

Author Contributions: The following statements should be used "Conceptualization, M.S., S.D., R.R. and F.L.; methodology, M.S., S.D., R.R. and F.L.; software, S.D.; validation, M.S. and S.D.; formal analysis, M.S., S.D. and R.R.; investigation, M.S.; data curation, M.S., S.D. and R.R.; writing—original draft preparation, M.S.; writing—review and editing, M.S., S.D., R.R. and F.L.; funding acquisition, R.R. All authors have read and agreed to the published version of the manuscript.

Funding: This research received no external funding.

Institutional Review Board Statement: Not applicable

Informed Consent Statement: Not applicable

Conflicts of Interest: The authors declare no conflict of interest.

References

1. Serpilli, M.; Rizzoni, R.; Lebon, F.; Dumont, S. An asymptotic derivation of a general imperfect interface law for linear multiphysics composites. *Int. J. Solids Struct* **2019**, *180-181*, 97–107.
2. Dumont, S.; Serpilli, M.; Rizzoni, R.; Lebon, F. Numerical Validation of Multiphysic Imperfect Interfaces Models. *Front. Mater.* **2020**, *7*, 158, doi: 10.3389/fmats.2020.00158
3. Kapitza P.L. *Collected Papers of P. L. Kapitza*; reprinted 1965; ter Haar, D., Ed.; Pergamon: Oxford, UK, 1941.
4. Benveniste Y. The effective conductivity of composites with imperfect thermal contact at constituent interfaces. *Int. J. Eng. Sci* **1986**, *24*,1537–1552.
5. Benveniste Y. Effective thermal-conductivity of composites with a thermal contact resistance between the constituents-nondilute case. *J. Appl. Phys.* **1987**, *61*, 2840–2843.
6. Pham Huy H.; Sanchez-Palencia, E. Phenomenes de transmission à travers des couches minces de conductivité elevée. *J. Math. Anal. Appl.* **1974**, *47*, 284–309.
7. Miloh, T.; Benveniste,Y. On the effective conductivity of composites with ellipsoidal inhomogeneities and highly conducting interfaces. *Proc. R. Soc. Lond. A* **1999**, *455*, 2687–2706.
8. Hashin, Z. Thin interphase/imperfect interface in conduction. *J. App. Phys.* **2001**, *89*, 2261–2267.
9. Javili, A.; Kaessmair, S.; Steinmann, P. General imperfect interfaces. *Comput. Methods Appl. Mech. Engrg.* **2014**, *275*, 76–97.
10. Benveniste, Y. Effective mechanical behaviour of composite materials with imperfect contact between the constituents. *Mech. Mat.* **1985**, *4*, 197–208.
11. Hashin, Z. Thin interphase/imperfect interface in elasticity with application to coated fiber composites. *J. Mech. Phys. Solids* **2002**, *50*, 2509–2537.
12. Gurtin, M.E.; Murdoch, A.I. A continuum theory of elastic material surfaces. *Arch. Ration. Mech. Anal.* **1975**, *57*, 291–323.
13. Yvonnet, J.; Le Quang, H.; He, Q.C. An XFEM/level set approach to modelling surface/interface effects and to computing the size-dependent effective properties of nanocomposites. *Comput. Mech.* **2008**, *42*, 119–131.
14. Yvonnet, J.; Mitrushchenkov, A.; Chambaud, G.; He, Q.C. Finite element model of ionic nanowires with size-dependent mechanical properties determined by ab initio calculations. *Comput. Methods Appl. Mech. Eng.* **2011**, *200*, 614–625.
15. Benveniste, Y.; Miloh, T. Imperfect soft and stiff interfaces in two-dimensional elasticity. *Mech. Mat.* **2001**, *33* 309–323.
16. Benveniste, Y. A general interface model for a three-dimensional curved thin anisotropic interphase between two anisotropic media. *J. Mech. Phys. Solids* **2006**, *54*, 708–734.
17. Ciarlet, P.G. *Mathematical Elasticity, Vol. II: Theory of Plates*; North-Holland: Amsterdam, The Netherlands, 1997.
18. Serpilli, M.; Lenci, S. Asymptotic modelling of the linear dynamics of laminated beams. *Int. J. Solids Struct.* **2012**, *49*, 1147–1157.
19. Serpilli, M.; Lenci, S. An overview of different asymptotic models for anisotropic three-layer plates with soft adhesive. *Int. J. Solids Struct.* **2016**, *81* 130–140.
20. Furtsev, A.; Itou, H.; Rudoy, E. Modeling of bonded elastic structures by a variational method: Theoretical analysis and numerical simulation. *Int. J. Solids Struct.* **2020**, *182–183*, 100–111.
21. Furtsev, A.; Rudoy, E. Variational approach to modeling soft and stiff interfaces in the Kirchhoff-Love theory of plates. *Int. J. Solids Struct.* **2020**, *202*, 562–574.
22. Klarbring, A. Derivation of the adhesively bonded joints by the asymptotic expansion method. *Int. J. Eng. Sci.* **1991** *29*, 493–512.
23. Geymonat, G.; Krasucki, F.; Lenci, S. Mathematical Analysis of a bonded joint with a soft thin adhesive. *Math. Mech. Solids* **1999**, *16*, 201–225.
24. Bessoud, A.L.; Krasucki, F.; Serpilli, M. Asymptotic analysis of shell-like inclusions with high rigidity. *J. Elast.* **2011**, *103*, 153–172.
25. Ljulj, M.; Tambaca, J. 3D structure – 2D plate interaction model. *Math. Mech. Solids* **2019**, *4*, 3354–3377.
26. Lebon, F.; Rizzoni, R. Asymptotic analysis of a thin interface: the case involving similar rigidity. *Int. J. Eng. Sci.* **2010**, *48*, 473–486.
27. Lebon, F.; Rizzoni, R. Asymptotic behavior of a hard thin linear interphase: An energy approach. *Int. J. Solids Struct.* **2011**, *48*, 441–449.
28. Rizzoni, R.; Dumont, S.; Lebon, F.; Sacco, E. Higher order model for soft and hard elastic interfaces. *Int. J. Solids Struct.* **2014**, *51*, 4137–4148.
29. Dumont, S.; Rizzoni, R.; Lebon, F.; Sacco, E. Soft and hard interface models for bonded elements, *Compos. Part B Eng.* **2018**, *153*, 480–490.
30. Lebon, F.; Rizzoni, R. Higher order interfacial effects for elastic waves in one dimensional phononic crystals via the Lagrange-Hamilton's principle. *Eur. J. Mech. A Solids* **2018**, *67*, 57–70.
31. Serpilli, M. Mathematical modeling of weak and strong piezoelectric interfaces. *J. Elast.* **2015**, *121*, 235–254.
32. Serpilli, M. Asymptotic interface models in magneto-electro-thermo-elastic composites. *Meccanica* **2017**, *52*, 1407–1424.
33. Serpilli, M. Classical and higher order interface conditions in poroelasticity. *Ann. Solid Struct. Mech.* **2019**, *11*, 1–10.
34. Serpilli, M. On modeling interfaces in linear micropolar composites. *Math. Mech. Solids* **2018**, *23*, 667–685.
35. Renard, Y.; Pommier, J. *Getfem++. An Open Source Generic C++ Library for Finite Element Methods*; Tech. rep.; INSA Lyon: Villeurbanne, France, 2002.
36. Geuzaine, C.; Remacle, J.-F. Gmsh: A 3-d finite element mesh generator with built-in pre- and post-processing facilities. *Int. J. Numer. Methods Eng.* **2009**, *79*, 1309–1331.

37. Shi, P.; Shillor, M. Existence of a solution to the N dimensional problem of thermoelastic contact. *Indiana Univ. Math. J.* **1992**, *17*, 1597–1618.
38. Khludnev, M.; Kovtunenko, V.A. *Analysis of Cracks in Solids*; WIT-Press: Southampton, UK; Boston, MA, USA, 2000.
39. Xu, X. The N-dimensional quasistatic problem of thermoelastic contact with Barbers heat exchange conditions. *Adv. Math. Sci. Appl.* **1996**, *6*, 559–587.
40. Showalter, R.E. Diffusion in poro-elastic media. *J. Math. Anal. Appl.* **2000**, *251*, 310–340.
41. Miara, B., Suarez, J.S. Asymptotic pyroelectricity and pyroelasticity in thermopiezoelectric plates. *Asympt. Anal.* **2013** *81*, 211–250.
42. Bonaldi, F.; Geymonat, G.; Krasucki, F.; Serpilli, M. An asymptotic plate model for magneto-electro-thermo-elastic sensors and actuators. *Math. Mech. Solids* **2017**, *22*, 798–822.
43. Jafarinezhad, M.R.; Eslami, M.R. Coupled thermoelasticity of FGM annular plate under lateral thermal shock. *Compos. Struct.* **2017**, *168*, 758–771.

MDPI
St. Alban-Anlage 66
4052 Basel
Switzerland
Tel. +41 61 683 77 34
Fax +41 61 302 89 18
www.mdpi.com

Technologies Editorial Office
E-mail: technologies@mdpi.com
www.mdpi.com/journal/technologies